8th International Conference on Advanced Materials, Mechanics and Structural Engineering (AMMSE 2021)

IOP Conference Series: Materials Science and Engineering
Volume 1222

Taipei, Taiwan
17 September 2021

ISBN: 978-1-7138-4680-2
ISSN: 1757-8981

Printed from e-media with permission by:

Curran Associates, Inc.
57 Morehouse Lane
Red Hook, NY 12571

Some format issues inherent in the e-media version may also appear in this print version.

Institute of Physics
Dirac House, Temple Back
Bristol BS1 6BE UK

Phone: 44 1 17 929 7481
Fax: 44 1 17 920 0979

techtracking@iop.org

Additional copies of this publication are available from:

Curran Associates, Inc.
57 Morehouse Lane
Red Hook, NY 12571 USA
Phone: 845-758-0400
Fax: 845-758-2633
Email: curran@proceedings.com
Web: www.proceedings.com

TABLE OF CONTENTS

Preface

The AMMSE 2021 is the 2021 8th International Conference on Advanced Materials, Mechanics and Structural Engineering (8th AMMSE 2021) was planned in Taipei, Taiwan during October 15-17, 2021. But currently, the entire world is struggling against the virulent pandemic COVID-19, so that the AMMSE 2021 conference changed the presentation method from Oral presentation to Remote Video Presentation to help with Global Quarantine Activities.

The Virtual of AMMSE 2021 was held on Incheon National University at Video Conference Meeting Room on October 15-17, 2021. The AMMSE Committee members attended conference at on-site and also authors attended the conference as Video Meeting. The conference program covered Invited Keynote Speech, Oral, and Poster Presentations, they presented their presentation 15 minutes including Q&A time.

The committee of AMMSE 2021 expresses their sincere thanks to all authors for their high-quality research papers and careful presentations. All reviewers are also thanked for their careful comments and advices. Thanks, are finally given to IOP Publication as well for producing this volume.

The Organizing Committee of AMMSE 2021

Committee Chair

Prof. Mosbeh Kaloop

Incheon Disaster Prevention Research Center

Incheon National University

IOP Conf. Series: Materials Science and Engineering 1222 (2022) 011001 doi:10.1088/1757-899X/1222/1/011001

Sponsors

Incheon Disaster Prevention Research Center (IDPRC)	IDPRC 인천대학교 방재연구센터 Incheon Disaster Prevention Center

Recently, various efforts to prevent and prepare are vitally needed for prevention of Disasters and calamities. So we understand the necessity for technology of disaster and we built up the Incheon Disaster Prevention Research Center (IDPRC) in Incheon National University (INU) in 1997.

Accordingly, Incheon Disaster Prevention Research Center (IDPRC) in Incheon National University has progressed of research on the prevention of disasters and calamities through the various seminars, conference and lectures. These research could be conducted cooperation with Incheon National University (INU) in various fields' structure, soil, hydraulic and environment.

Incheon Disaster Prevention Research Center (IDPRC) will try to be a leader in the disaster of industry through the various research activities and global conference.

IOP Conf. Series: Materials Science and Engineering 1222 (2022) 011001 doi:10.1088/1757-899X/1222/1/011001

Lists of Committees

Committee name	Organizer name	Organization
Honorary Chair	Mosbeh Kaloop	Incheon National University
Conference General Chairs	Mosbeh Kaloop	Incheon National University
Honorary Chair	Dongkeon Kim	Dong-A University
Technical Committees	J. W. Hu	Incheon National University

IOP Conf. Series: Materials Science and Engineering 1222 (2022) 011002 doi:10.1088/1757-899X/1222/1/011002

Peer Review Statement

All papers published in this volume have been reviewed through processes administered by the Editors. Reviews were conducted by expert referees to the professional and scientific standards expected of a proceedings journal published by IOP Publishing.

- **Type of peer review:** Double Anonymous
- **Conference submission management system:** Morressier
- **Number of submissions received:** 48
- **Number of submissions sent for review:** 36
- **Number of submissions accepted:** 17
- **Acceptance Rate (Submissions Accepted / Submissions Received × 100):** 35.4
- **Average number of reviews per paper:** 3
- **Total number of reviewers involved:** 3
- **Contact person for queries:**
  Name: R. Kaloop Mosbeh
  Email: mosbeh@inu.ac.kr
  Affiliation: Incheon Disaster Prevention Research Center

IOP Conf. Series: Materials Science and Engineering 1222 (2022) 012001 doi:10.1088/1757-899X/1222/1/012001

Combined Subset Simulation and Comprehensive Learning Particle Swarm Optimization in Reliability-Based Structural Optimization

Arnut Sutha[1,a], Thu Huynh Van[1,b], Sawekchai Tangaramvong[1,c]

[1]Applied Mechanics and Structures Research Unit, Department of Civil Engineering, Chulalongkorn University, 10330 Bangkok, Thailand

[a]6272105521@student.chula.ac.th, [b]6278306421@student.chula.ac.th
[c]sawekchai.t@chula.ac.th

Abstract. Reliability-based design optimization (RBDO) addresses the cost-effective integrity design of structures in the presence of inherent uncertain parameters. Processing this class of problem is challenging from the computational burden to determine the failure probability of structures violating the limit-state function. This paper proposes an efficient decoupling RBDO method that advantageously couples a comprehensive learning particle swarm optimization (CLPSO) algorithm with a subset simulation (SS), termed as SS-CLPSO approach. In essence, the proposed method iteratively performs the CLPSO assuming deterministic parameters based on the most probable point underpinning limit-state functions updated within the reliability evaluation process. Based on the CLPSO design data, the SS approximates the spectrum of limit-state functions under uncertain parameters, and hence enables the significant reduction of Monte-Carlo simulations for the failure probability prediction. The SS map outs the failure probability from the conditional samples constructed at each intermediate event. The proposed SS-CLPSO terminates the optimal solution to the RBDO problem as when the resulting failure probability converges to the permissible threshold. The applications of the present approach are illustrated through the steel truss design under probabilistic uncertain parameters and constraints.

1. Introduction

Deterministic optimization has been extensively applied in engineering structures to improve the design performance with minimum resources. The design solution computed by the deterministic optimization becomes unreliable in some cases, especially when the influences of uncertainties inheriting structural dimensions, material properties, loading and operating conditions are significant and cannot be eliminated. By addressing the performance and reliability of the structure together, the structural reliability-based design optimization (RBDO) has been considered as the alternative approach in recent years. More explicitly, the RBDO problem minimizes the cost function, denoted as C, and satisfies the certain deterministic and probabilistic constraints, as state by the following generic mathematical formulations [1]:

$$\begin{aligned} \mathbf{min}\ \ & C(s) \\ s.t.\ \ & \mathbb{P}\left[G(s,x)-\overline{z}\leq 0\right]-P_a\leq 0 \\ & s\in\left[s_L,\ s_U\right] \end{aligned} \quad (1)$$

where **s** and **x** are the vectors of deterministic design variables and random parameters, respectively. The vector **x** is characterized by the joint probability density function (PDF) $f(\mathbf{x})$ in the space Ω. The two s_L and s_U denote the lower and upper bounds on the variables **s**. The functions $G(\mathbf{s},\mathbf{x})$ respectively the performance and limit-state design expressions considered, where $\overline{z}$ is a constant threshold of the failure domain specified for the design variables (i.e. $\mathrm{F}=\left\{\boldsymbol{x}\in\Omega\,|\,G(\boldsymbol{s},\boldsymbol{x})-\overline{z}\le 0\right\}$. The probability of failure reads $P_f=\Pi\left[\mathrm{G}(\mathrm{s},\ \mathrm{x})-\overline{\mathrm{z}}\ \le 0\right]=\int_{\mathrm{F}} f(\boldsymbol{x})\mathrm{dx}$, where P_a is the allowable threshold of P_f.

2. CLPSO Algorithm

Whilst the general PSO algorithm provides the good convergence rate to the optimal solutions, the method is often trapped into the local optimal. To enable the diversity of design particles and overcome the premature convergence, Liang and Qin [2] developed the CLPSO algorithm using the comprehensive learning (CL) strategy. The position $X_i^{d,\mathrm{next}}$ and velocity $V_i^{d,\mathrm{next}}$ functions of CLPSO are written as follows:

$$X_i^{d,next}=X_i^d+V_i^{d,next} \tag{2}$$

$$V_i^{d,next}=wV_i^d+cr_1\left(pbest_{f_i(d)}-X_i^d\right) \tag{3}$$

Where i and d are the array indices of the particles for $i\in\{1,...,ps\}$ and dimensions for $d\in\{1,...,nd\}$, respectively; X_i^d and V_i^d the position and velocity of the i-th particle at the d-th dimension, respectively; w the inertia weight; c the acceleration coefficient (equal to 1.5) attracting X_i^d to $X_i^{d,\mathrm{next}}$; r_1 a random number in the range of $[0,1]$; f_i the exemplar index the i-th particle follows; $pbest_{f_i(d)}$ the best position of the f_i-th particle for the d-th dimension. The updates for X_i^d and V_i^d across swarm population comply with the learning probability function [3]:

$$Pc_i=0.05+0.45\left(\mathbf{exp}\left(\frac{10(i-1)}{N_p-1}\right)-1\right)/\left(\mathbf{exp}(10)-1\right) \tag{4}$$

3. Subset Simulation

Subset simulation (SS) is a well-known, efficient Monte Carlo technique for variance reduction in the structural reliability community. It exploits the concept of conditional probability and advanced Markov Chain Monte Carlo technique [4]. The formulation for SS optimization is given by

$$P_f=P(F)=P(F_1)\prod_{k=1}^{m-1}P(F_{k+1}\,|\,F_k) \tag{5}$$

where F_k for $\forall k\in\{1,...,m\}$ are the intermediate failure events nested (satisfying $F_1\supset F_2\supset...\supset F_m=F$). The nesting feature of all intermediate events gives the decomposition of small probability. Then, searching for C(s) in Eq. (1) is converted to the failure region in the reliability problem. A key step for the success of SS is the computation of conditional samples for each intermediate event to estimate its corresponding failure probability. Because the probability density functions (PDFs) for intermediate events are implicit functions, it is not practical to generate samples directly from their PDFs. This can be achieved using Markov Chain Monte Carlo (MCMC). A modified Metropolis–Hastings (MMH) algorithm has been developed for SS and summarized.

Modified Metropolis-Hastings algorithm

Let $\mathbf{s}\sim\pi(.\,|\,F^1)$ be a sample from the conditional distribution $\pi(.\,|\,F^1)$. The MMH algorithm generates another sample $\overline{\mathbf{s}}$ from $\pi(.\,|\,F^1)$ as follows [5]:

1. Generate a "candidate" sample ξ : for each coordinate j = 1, …, nd
   a) Sample $\eta_j \sim q_j(\cdot|\mathbf{s}_j)$, where $q_j(\cdot|\mathbf{s}_j)$,called the proposal distribution, is a univariate PDF for η_j centered at $\mathbf{s}_j$ with the symmetry property $q_j(\eta_j|\mathbf{s}_j)=q_j(\mathbf{s}_j|\eta_j)$. The proposal distribution adopts a Gaussian PDF with mean $\mathbf{s}_j$ and variance σ_j^2:

$$q_j(\eta_j|s_j)=\frac{1}{\sqrt{2\pi}\sigma_j}\mathbf{exp}\left(-\frac{(\eta_j-s_j)^2}{2\sigma_j^2}\right) \quad (6)$$

   b) Define the jth coordinate of the candidate sample by accepting or rejecting η_j:

$$\xi_j=\begin{cases}\eta_j\text{ ,with probability } \mathbf{min}\left\{1,\frac{\O(\eta_j)}{\O(s_j)}\right\}\\ s_j\text{, otherwise}\end{cases}$$, where Ø (·) denotes the PDF. 7)

2. Accept or reject the candidate sample ξ by setting

$$\tilde{s}=\begin{cases}\xi\ , & if\ \xi\in F_1\\ s\ , & if\ \xi\notin F_1\end{cases} \quad (8)$$

4. Failure Probability Estimation Using MCS

Monte-Carlo simulations can be performed to approximate the failure probability in Eq. (1) by

$$P_f=\int_\Omega f(x)dx=\frac{1}{N_{MCS}}\sum_{j=1}^{N_{MCS}}\left[G_j(s,x_j)-\overline{z}\right] \quad (9)$$

where N_{MCS} is the number of random samples generated within the space Ω, and [.] denotes the indicator function, namely

$$\left[G_j(s,x_j)-\overline{z}\right]=\begin{cases}1 & if\quad G_j(s,x_j)-\overline{z}\le 0\\ 0 & otherwise\end{cases} \quad (10)$$

5. Combined SS-CLPSO Algorithm

Th proposed SS-CLPSO method for the optimal solution of the RBDO in Eq. (1) is summarized:

Step 1: Initialize the random variables in the original RBDO Eq. (1).

Step 2: Perform the deterministic optimization using the CLPSO algorithm to obtain the optimal design solutions.

Step 3: Import $\mathbf{s}_j$ into SS using $\mathbf{s}_j+\mathcal{N}(0,1)$, number of n_c samples and n_s cycles, and then sort the results G_0(**s**, **x**) in descending order first. Store the value $G_0^{(nc)}$ (**s**, **x**), then select n_c sample to create the next level.

Step 4: The new sample group must be less than $G_0^{(\mathrm{nc})}$ (**s**, **x**), the number of n_c samples and n_s cycles, and then sort the results G_1(**s**, **x**) in ascending order first. Collect the value $G_1^{(nc)}$ (**s**, **x**), then select n_c for example and calculate P_f, If $P_f > P_a$, repeat this Step 4 until x_{MMP} is achieved.

Step 5: Bring x_{MMP} repeat Steps 2 to 5. If the estimated failure probability P_f converges, terminate, where $n_c = np$ is the number of Markov chains, $n_s = n/n_c$ is the number of states in each chain, n is the number of samples per level, and p= [0.1,0.2] is the level probability

6. Illustrative Example

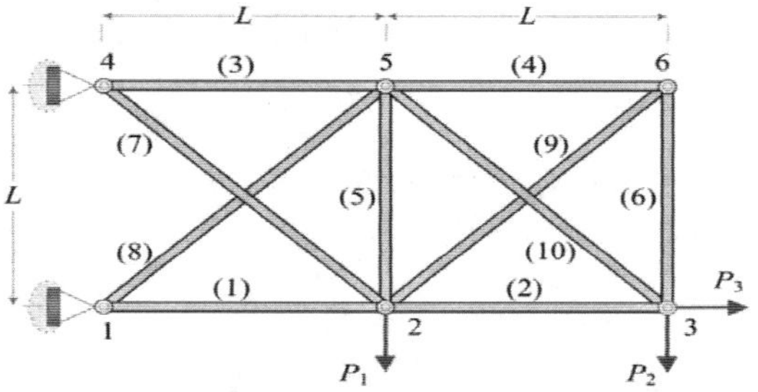

Figure 1. Ten-bar truss.

Table 1. Probabilistic properties of random parameters.

Variable	Distribution	Mean value	COV
P_1 [kN]	Normal	60	0.20
P_2 [kN]	Normal	40	0.20
P_3 [kN]	Normal	10	0.20
E [GPa]	Normal	200	0.10
L [m]	Normal	1	0.05

A 10-bar truss in Fig. 1 was considered [1]. The RBDO problem in Eq. (1) was formulated. The cross-sectional areas were employed the design variables, namely $s=\left[s_1,\ldots,s_{10}\right]^T$, where each variable is bounded within the intervals, i.e., $s_i \in\left[1,20\right]\times 10^{-4}\, m^2.$. The intrinsic (indeterministic probabilistic) uncertainties inherited the external loads (P_1, P_2 and P_3), Young's modulus (E) and dimension length (L), simultaneously. The random variables were $x=\left[P_1,P_2,P_3,E,L\right]^T$ with the probabilistic properties listed in Table 1. The vertical displacement at node 3, denoted as Δ_3, was considered as the response performance of interest ($y=-\Delta_3$), whose probability of exceeding the allowable value of $\overline{z}=4\times 10^{-3}$ m. was less than or equal to P_a=6.21×10^{-3}. Thus, the specific RBDO Eq. (1) was written as follows:

$$\begin{aligned} &\min C(s)=\sum_{i=1}^{10} s_i \\ &s.t.\quad \mathbb{P}\left[G(s,x)+4\times 10^{-3}\le 0\right]-6.21\times 10^{-3}\le 0 \\ &\qquad s_i\in\left[1,20\right]\times 10^{-4}\left(m^2.\right)\ ,i=1,\ldots,10 \\ &\qquad G(s,x)=-\Delta_3 \end{aligned} \tag{11}$$

In the first instance, the initial random variables were assigned to take the mean values given in Table 1. The deterministic counterpart of the problem in Eq. (11) was solved using the CLPSO algorithm. The parameters adopted: the total number of particle populations of ps = 30; c =1.5; the inertial weight of w linearly declining from 0.9 to 0.4; and the maximum number of iterations of 1,000 as per each particle set. The parameters employed in SS were samples per level n=1,000 and level probability p =0.1 set.

The coupling CLPSO and SS procedures were iterated until the estimated probability of failure (viz., P_f = 6.07×10^{-3}) was converged and well complied with the limit of P_a = 6.21×10^{-3}. The proposed SS-CLPSO method was encoded in Python, and run using the computer hardware with RYZEN 7 4800 HS CPU @ 2.9 GHz and 16 GB RAM. The optimal solutions reported in Table 2 were successfully computed and agreed very well with the benchmark references [1, 6, 7].

Table 2. Comparisons of optimization results for various analysis approaches.

Design variables	FEA -FORM [6]	RSM2-FORM-MCS [6]	PSA-ISAP [7]	MGP-SA [1]	GRP-CLPSO [1]	SS-CLPSO (This Study)
s_1	10.493	10.705	10.482	10.333	10.635	9.924
s_2	5.772	5.914	4.421	5.371	5.589	5.750
s_3	14.098	14.424	15.685	13.579	13.481	13.845
s_4	1.000	1.000	1.089	1.000	1.000	1.082

s_5	1.000	1.000	1.000	1.000	1.000	1.030
s_6	1.000	1.000	1.000	1.000	1.000	1.006
s_7	5.460	5.531	7.851	6.418	5.883	5.913
s_8	11.586	11.853	10.048	11.273	11.149	11.165
s_9	1.000	1.000	1.121	1.000	1.000	1.042
s_{10}	10.958	11.223	9.650	10.508	10.555	10.186
$C(\mathbf{s})$ [$\times 10^{-4}$ m^2]	62.367	63.649	62.347	61.482	61.293	60.943
No. of FEA iterates	2,240	1,904	524	20,000	225 +15	4 Cycles
Computational time [s]	-	-	-	3,208	2,213	-
$\mathbb{P}[G(\mathbf{s}, \mathbf{x}) - \overline{z} \leq 0]$	8.51×10^{-3}	6.19×10^{-3}	6.19×10^{-3}	4.34×10^{-3}	5.43×10^{-3}	2.98×10^{-3}
P_f(MCS 10^6 samples)	3.59×10^{-3}	5.86×10^{-3}	5.72×10^{-3}	5.25×10^{-3}	5.22×10^{-3}	6.07×10^{-3}

Note: *4 cycles =10,000^{SS} samples. +4,000^{CLPSO} iterates.*

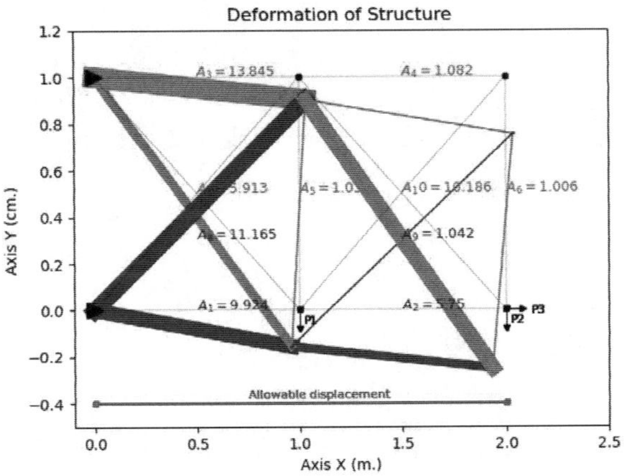

Figure 2. Normalized cross-sectional areas of optimally designed members, where blue and red-colour bars indicate compression and tension members, respectively.

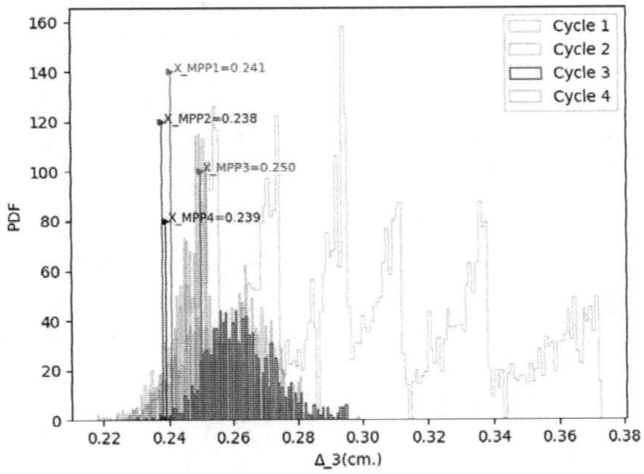

Figure 3. Sample generated by SS, where red sample are MCMC generated at conditional level convergence to P_a and red line is the most probable point underpinning the limit-state function

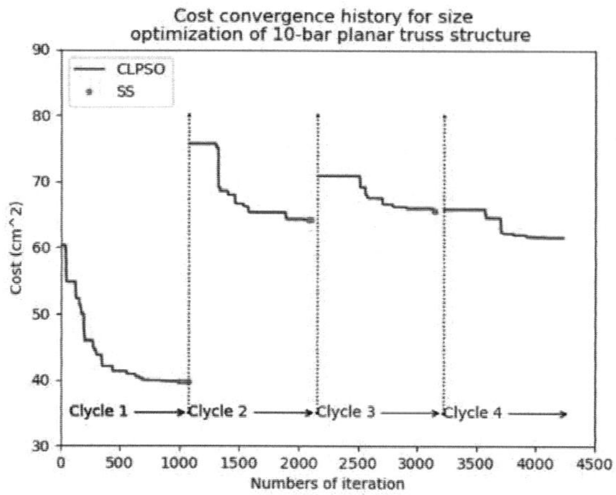

Figure 4. Solution convergence computed by SS-CLPSO.

7. Conclusion

The combined SS-CLPSO method determine the optimal solution of the RBDO problem. It is based the cooperative procedures between CLPSO (the sizing design of members) and SS (the efficient failure probability approximations). A number of steel truss design examples and benchmarks (one of which has been provided herein) illustrate the robustness and accuracy of the proposed RBDO solution approach.

Acknowledgments

This research has been supported under
Thailand Research Fund (Grant No.6280089).

References

[1] Van TH, Do B, Limkatanyu S, Tangaramvong S. Combined Gaussian process regression model and comprehensive learning particle swarm optimization in reliability-based structural optimization. Proceedings, the 12th International Conference on Computational Methods (July 4-8, 2021 Online). 2021:335-42.

[2] Liang JJ, Qin AK, Suganthan PN, Baskar S. Comprehensive learning particle swarm optimizer for global optimization of multimodal functions. IEEE Transactions on Evolutionary Computation. 2006;10:281-95.

[3] Zhang K, Huang Q, Zhang Y. Enhancing comprehensive learning particle swarm optimization with local optima topology. Information Sciences. 2019;471:1-18.

[4] Au, S.K.; Beck, J.L. Estimation of small failure probabilities in high dimensions by subset simulation. Probabilistic Engineering Mechanics 2001, 16, 263–277.

[5] Konstantin Zuev. Subset Simulation Method for Rare Event Estimation: An Introduction. Springer, Berlin, Heidelberg. 2015

[6] Zhao W, Qiu Z. An efficient response surface method and its application to structural reliability and reliability-basedoptimization. Finite Elements in Analysis Design. 2013;67:34-42.

[7] Ghasemi MR, Camp CV, Dizangian B. Novel decoupled framework for reliability-based design optimization of structures using a robust shifting technique. Frontiers of Structural and Civil Engineering. 2019;13:800-20.

Determining the Optimum Geometric Shape for Web Opening of Wide Flange Beams using Response Surface Methodology

K R R Padunan[1] and G B Estores[2]

Muralla St., Intramuros, Manila, Philippines

krrpadunan@mapua.edu.ph[1]; gbestores@mapua.edu.ph[2]

Abstract. In the field of civil engineering, the priority of a structural engineer is the safeness and stability of each structural element present in the structure. However, there were some concerns with regards to easy access and flexibility of the utilities such as electrical and mechanical systems. In response to these concerns, web opening was introduced to wide flanges to give way for the utilities, provided that the strength and durability of the structure will not be compensated. This study aims to obtain the optimum geometric shape of web opening for wide flange beams with regards to its performance for bending stress. The web opening on the wide flanges was designed first based on the parameters of the web opening. Then, the effects of circular, square, hexagonal and elliptical web openings were analyzed using Finite Element Analysis (FEA). Three varying support conditions in each opening case were used and analyzed using nonlinear static and nonlinear dynamic analysis in the finite element modeling software called Abaqus. During the FEA, deflection control method was used to obtain the value of force and stress upon reaching the deflection control value. Values of maximum stress and maximum force for each case were then obtained from the analysis, were used as factors for the Response Surface Methodology (RSM), which will define which shapes performed the best under specific conditions. Results obtained from RSM showed that considering value of the stress and deflection from nonlinear static and nonlinear dynamic analysis, elliptical web opening produced the largest force and smallest stress. Thus, elliptical web opening, considering bending stress, performed better compared to the other web openings.

Keywords: Web Opening, Wide Flange Beams, Finite Element Analysis, Abaqus, Response Surface Methodology

1. Introduction

Way back, being a structural engineer normally means utilizing solid sections to be used as its structural elements such as beams and girders in order to erect steel structures. Having said that, it is their responsibility to address the safeness of the structure that was built and designed and together with it, is its serviceability. However, having the two the main objective of the structural engineer in designing. It should not be forgotten that the designer also needs to consider on how to maximize the functionality of the structural system which pertains to the usage of and other uses of its structural elements. To comply with it, adjustments have been made like increasing the floor height and others had the idea of punching a hole through the web in order to give way for additional way of utilities. Web openings have been used in many structures especially the ones made with steel sections to comply with the functionality of the structural steel element. Commonly, the web openings of steel beams are created to improve the function of structures such as for air-conditioning ducting, electrical conduits, water supply and sanitary pipelines. Way back, web openings were created by punching the opening right at the location. However, its shows that having web openings changes the stress distribution and collapse behavior. among the structural elements especially the beams and girders. Having web openings will eventually take an effect on capacity of the beam to carry load [1]

In a competitive world, with increasing cost for materials and labor, the search for the safest design economy consistent with safety and the desired life of the structure is of major importance. Members must be shaped, arranged and connected in ways that it will provide safety, economical and efficient solution to problems. Structural safety is attained through the right design, good fabrication workmanship and methods. The avoidance of any possibility of structural failure should be primary concern of the designer. Years if design experience, conditioned by both unsuccessful and successful behavior, have produced criteria that aid the choice of safe stress levels. Theses may not always produce the most economical structure; nevertheless, the overall cumulative experience in engineering design has provided designers with confidence and also let them pursue attempts to a new venturesome type of structures or members. [2]

The practice of having web openings in steel beams has been taken into consideration especially the effect of the opening on the design criteria or aspect and the capacity itself of the beam. Due to this reason, there are countries that established their national structural standards pertaining to having web openings in beams. These rules tackle the concern regarding the weakening of the beam due to web openings. However, based on what the rules shows, the rules implemented does not make the design conservative but if more studies will be done and if the parameters of the design will be adjusted, having web openings will be a lot efficient as it will become more flexible and increased economical effect. [3]

The finite element method is a numerical procedure which can model the behavior of a structure w ith great accuracy. Finite element modelling is normally used to deal with complicated one-dimensional, two-dimensional and three-dimensional (1D,2D,3D) geometry together with the variation among its material properties and some structural restrains. It is done by reducing the governing partial differential equations for the structure into systems of linear algebraic equations. It serves as an alternative to dealing directly with the governing partial differential equations to obtain algebraic equation systems which is then used to treat the structure in terms of its energy and work done on it. The understanding and application of the finite element method to structural system requires designers to have an idea on the vat number of concepts regarding mathematics and structural design. However, it is not needed to be really good at it or have a mastery pertaining to it but just having a knowledge regarding the limitations and good manipulative skill in the utilization and application of theorems and procedures will do a lot in the application of those in the finite element method. [4]

Response Surface Methodology (RSM) is commonly described as technique which utilizes various complex calculation techniques which helps in the optimization of the response or the output variable. This method is most likely developed in order to be suitable and compatible in creating various experimental design that will incorporate all of the independent variable that will be utilized in the modelling and also this technique utilizes all of the input data in order to produce various equations which will yield outputs based on theoretical value. Using the response surface methodology is not just by combining variables, it provides parameters for the independent variables to yield controlled variables obtained from the regression analysis. [5]

2. Methods

2.1. Design Parameters for Finite Element Analysis.
For this study, certain modes of failure is to be considered first before anything else. Three modes of failure were considered in choosing a section based on American Institute for Steel Construction (AISC) design parameters.

The chosen section should be able to pass against the following:
Local buckling

$$\frac{b_f}{2t_f} \leq \frac{65}{\sqrt{F_y}} \quad (1)$$

Web buckling

$$p_o = \frac{a_o}{h_o} + \frac{6h_o}{d} \quad (2)$$

Lateral buckling

$$[1-(\frac{a_o}{L_b}) + \frac{\Delta A_s}{t_w(d+2b_f)}]^2 \leq 1 \quad (3)$$

After selecting the section based on the given mode of failures mentioned, design for web opening will be done by following these certain equations:

For concentrated loads, no loads should be placed above web openings unless needed otherwise bearing stiffeners are not required to prevent web crippling in the vicinity of an opening due to concentrated load if

$$\frac{d-2t_f}{t_w} \leq \frac{420}{\sqrt{F_y}} \quad ; \quad \frac{b}{t} \leq \frac{54}{\sqrt{F_y}} \quad (4)$$

and the load should be placed d/2 from the edge of the opening or

$$\frac{d-2t_f}{t_w} \leq \frac{520}{\sqrt{F_y}} \quad ; \quad \frac{b}{t} \leq \frac{65}{\sqrt{F_y}} \quad (5)$$

and the load should be placed d/2 from the edge of the opening and in any case the edge of an opening should not be closer than a distance d to a support
For circular openings
$h_o = D_o$ for bending
$h_o = 0.9 D_o$ for shear
$a_o = 0.45 D_o$
D_o is the diameter of the opening
Spacing of openings
Rectangular openings

$$s \geq h_o \quad (6)$$

$$s \geq a_o \left(\frac{V_u/\emptyset V_p}{1 - V_u/\emptyset V_p}\right) \quad (7)$$

Circular Openings

$$s \geq 1.5 D_o \quad (8)$$

$$s \geq D_o\left(\frac{V_u/\emptyset V_p}{1 - V_u/\emptyset V_p}\right) \quad (9)$$

wherein S = clear space between openings

2.2. Geometry of the Steel Beams and its opening

Twelve cases have been determined for this study, but before getting into the detail of those cases, it is more important to determine which section of steel beam will be used for the modelling and the geometry of the web openings for the model. In order to determine which steel section is appropriate for the study, it needs first to be based from the American Institute of Steel Construction which pertains to steel and composite beams with web opening. The sectional properties needed first to be evaluated against local buckling, web buckling, and lateral buckling. For the many sections that passed the evaluation, W12x170 (A35) was chosen for this study and the following are the properties of the said section.

Table 1. Steel Sectional Properties

Section	*Area*	*Depth*	*Web*		*Flange*	
	A	*d*	*tw*	*tw/2*	*bf*	*tf*
	sq.mm	*mm*	*mm*	*mm*	*mm*	*mm*
W12x170	32258	356.36	24.38	12.19	319.28	39.62

For the modelling of the beams, the following figures below shows the typical layout for the varying support condition of each of the cases.

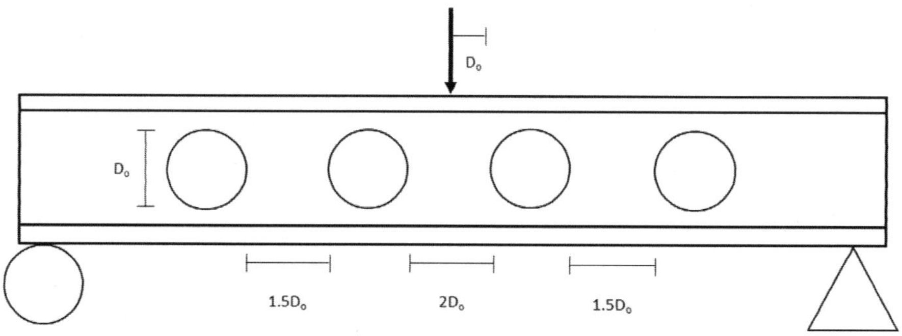

Figure. 1 Simply Supported Beam Typical Design Layout

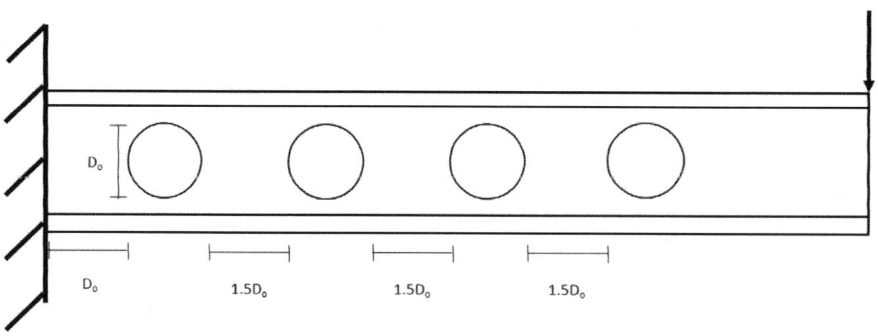

Figure. 2 Cantilever Beam Typical Design Layout

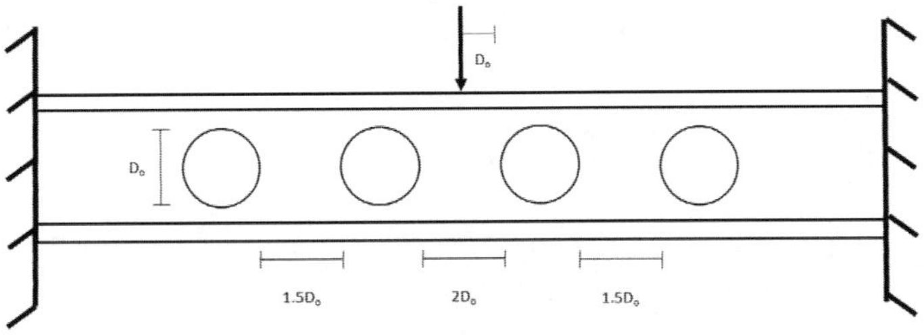

Figure. 3 Fully Restrained Beam Typical Design Layout

In this study, it has two ideal concepts for the web opening, one is that the opening will be based on the code, and the other one is where utilities will pass through such that it is based for the 100 mm diameter for the web opening.

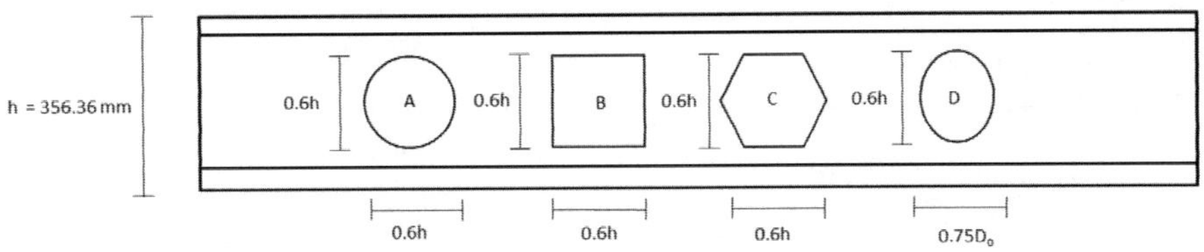

Figure. 4 Proportion of Geometric Opening to the height of the beam

Table 2. Geometry of the Web Opening based in accordance with the AISC

Opening Parameters as predetermined in the code		
Shape	*Ao (Length of the opening)*	*ho (height of the opening)*
Circular	213.816	213.816
Square	213.816	213.816
Regular Hexagon	213.816	213.816
Vertical Ellipse	160.362	213.816

Table 3. Geometry of the Web Opening determined by the 100 mm Diameter

Opening Parameters determined by the 100 mm diameter		
Shape	*Ao (Length of the opening)*	*ho (height of the opening)*
Circular	100	100
Square	100	100
Regular Hexagon	100	100
Vertical Ellipse	100	134

The following set of data shown on table 2 and 3 will be the one used in creating the finite element model to be analyzed in Abaqus. Results for the maximum stress and maximum deflection under deflection control method will be obtained through nonlinear static analysis and nonlinear dynamic analysis.

3. Results and Discussion

As the results were gathered, variables and responses can now be obtained in order to produce the maximum response the shape can yield. The variables to be considered in the study will be the independent variables and the dependent variables which are both essential in producing an optimum response using Response Surface Methodology. For the independent variable, there will be two variables, which is the length of the opening and the total area of the opening caused by the hole. For the dependent variable, the one to be evaluated is the result for the maximum stress and maximum force from the analyses.

Table 4. Result of Force and Stresses for the Circular Web Opening

			Independent Variable		*Response*	
AISC PARAMETERS		*Shape*	*Ao (length of opening) (m)*	*Total area of opening (m^2)*	*Force (N)*	*Stress (MPa)*
Static	Simply Supported	Circular	0.213816	0.143625	1589003	370.9
	Cantilever Beam	Circular	0.213816	0.143625	252493	415.4
	Fully Restrained	Circular	0.213816	0.143625	1904296	393.3
Dynamic	Simply Supported	Circular	0.213816	0.143625	1557441	363.5
	Cantilever Beam	Circular	0.213816	0.143625	329396	304.4
	Fully Restrained	Circular	0.213816	0.143625	1942413	384.9
100 mm basis						
Static	Simply Supported	Circular	0.1	0.00785	1910585.8	348.7
	Cantilever Beam	Circular	0.1	0.00785	339174.9	292.3
	Fully Restrained	Circular	0.1	0.00785	2575509.5	354.3
Dynamic	Simply Supported	Circular	0.1	0.00785	1969052	365.5
	Cantilever Beam	Circular	0.1	0.00785	336536	308.0
	Fully Restrained	Circular	0.1	0.00785	2568495	354.7

Table 5. Result of Force and Stresses for the Square Web Opening

			Independent Variable		*Response*	
AISC PARAMETERS		*Shape*	*Ao (length of opening) (m)*	*Total area of opening (m^2)*	*Force (N)*	*Stress (MPa)*
Static	Simply Supported	Square	0.213816	0.182869	1156507.3	334.2
	Cantilever Beam	Square	0.213816	0.182869	244244.7	405.5
	Fully Restrained	Square	0.213816	0.182869	1423344	399
Dynamic	Simply Supported	Square	0.213816	0.182869	1186745	333
	Cantilever Beam	Square	0.213816	0.182869	372546.5	292.7
	Fully Restrained	Square	0.213816	0.182869	1572831	394.2
100 mm basis						
Static	Simply Supported	Square	0.1	0.01	1909393.8	337
	Cantilever Beam	Square	0.1	0.01	338377.3	290.8
	Fully Restrained	Square	0.1	0.01	2563676.5	347.4
Dynamic	Simply Supported	Square	0.1	0.01	1964636	350.4
	Cantilever Beam	Square	0.1	0.01	318748.8	306.6
	Fully Restrained	Square	0.1	0.01	2541516	347.4

Table 6. Result of Force and Stresses for the Hexagon Web Opening

			Independent Variable		*Response*	
AISC PARAMETERS		*Shape*	*Ao (length of opening) (m)*	*Total area of opening (m^2)*	*Force (N)*	*Stress (MPa)*
Static	Simply Supported	Hexagon	0.213816	0.16	1463218.3	365.2
	Cantilever Beam	Hexagon	0.213816	0.16	331673.8	303.7
	Fully Restrained	Hexagon	0.213816	0.16	1797450	386.5
Dynamic	Simply Supported	Hexagon	0.213816	0.16	1483556	363.5
	Cantilever Beam	Hexagon	0.213816	0.16	388158.8	303.7
	Fully Restrained	Hexagon	0.213816	0.16	1815760	382.5
100 mm basis						
Static	Simply Supported	Hexagon	0.1	0.00866	1898544.4	346.9
	Cantilever Beam	Hexagon	0.1	0.00866	339055.9	293
	Fully Restrained	Hexagon	0.1	0.00866	2567842.8	349.4
Dynamic	Simply Supported	Hexagon	0.1	0.00866	1957815	355.9
	Cantilever Beam	Hexagon	0.1	0.00866	318164.9	309
	Fully Restrained	Hexagon	0.1	0.00866	2546436	349

Table 7. Result of Force and Stresses for the Ellipse Web Opening

			Independent Variable		*Response*	
AISC PARAMETERS		*Shape*	*Ao (length of opening) (m)*	*Total area of opening (m^2)*	*Force (N)*	*Stress (MPa)*
Static	Simply Supported	Ellipse	0.160362	0.215437	1679468.6	356.6
	Cantilever Beam	Ellipse	0.160362	0.215437	335962.5	290
	Fully Restrained	Ellipse	0.160362	0.215437	2123735	379.7
Dynamic	Simply Supported	Ellipse	0.160362	0.215437	1692968	340.8
	Cantilever Beam	Ellipse	0.160362	0.215437	329954.4	306.1
	Fully Restrained	Ellipse	0.160362	0.215437	2143256	375.7
100 mm basis						
Static	Simply Supported	Ellipse	0.1005	0.084615657	1860529.1	340.1
	Cantilever Beam	Ellipse	0.1005	0.084615657	338548.1	292.3
	Fully Restrained	Ellipse	0.1005	0.084615657	2487999.8	356.6
Dynamic	Simply Supported	Ellipse	0.1005	0.084615657	1924184	353.7
	Cantilever Beam	Ellipse	0.1005	0.084615657	355395.7	309.2
	Fully Restrained	Ellipse	0.1005	0.084615657	2509804	355.3

The utilization of the response surface methodology helped the researcher determine which geometric section performed the best under bending stress through nonlinear static and nonlinear dynamic analysis considering that the beam with varying web opening will also have different support conditions. Thus, by inputting the values on tables 4, 5, 6, and 7 will yield the value needed to evaluate the result of the performance of the geometric web openings given.

Upon simulation of the tables in Minitab, the following values may be obtained

Table 8. Summary of Optimization Results for Force and Stress

Summary of Optimization Results		
Shape	*Force (kN)*	*Stress (Mpa)*
Circular	1262.507	372.0667
Square	1327.26	343.4939
Regular Hexagon	1430.714	341.4148
Vertical Ellipse	1579.515	334.5333

4. Conclusion

The primary objective of this study is to explore the effect of having varying geometric shape for the web openings of wide flange beams and determine which will yield the optimum performance for the bending stress. This may be obtained by subjecting the wide flange beams with different geometric shapes given that the beam will also be under varying support conditions. The beam will then be subjected to nonlinear static analysis and nonlinear dynamic analysis wherein the maximum stress and maximum force will be obtained through the use of the deflection control method wherein the stress and force experienced by the section will be obtained when the deflection limit is attained. Values of the length of opening and total area of opening will then be used as independent variables while the values of the maximum force and maximum stress will serve as the response to the RSM. With those procedures, it will then yield that the ellipse performed the best under given conditions such that upon optimization, the ellipse yielded the highest value of force, which is 1579.515 kN, to attain the limit for deflection control and the same time, it shows that it also has the least stress experienced, which is 334.5333 MPa compared to the other geometric shapes that were given. These results only conclude that the remaining area on the web does not determine the overall capacity of the beam, thus it depends on the geometric shape that was used for the web opening.

IOP Conf. Series: Materials Science and Engineering 1222 (2022) 012002 doi:10.1088/1757-899X/1222/1/012002

References

[1] El-Dehemy, H. (2017). Static and Dynamic Analysis Web Opening of Steel Beams .
[2] Darwin, D. e. (N.D.). Composite beams with web openings.
[3] Jaiswal, A. e. (2016). Study of Steel Beam with Web Openings: A Review .
[4] Kumar, U. (April 2018). A study on flexural capacity of steel beams with corrugated web
[5] Amin, M. e. (2015). Overview on the Response Surface Methodology (RSM) in Extraction Processes.

doi:10.1088/1757-899X/1222/1/012003

Finite Element Study on Forming Helical Springs by Wire Bending with a Plate

K J Fann

Department of Mechanical Engineering, National Chung Hsing University, 250, Kuokuang Road, Taichung 40227, Taiwan, R.O.C

kjfann@nchu.edu.tw

Abstract. Coil springs, widely used as mechanical elements, will enter another new production outlook due to the application of CNC wire bending machines and as the era of On-Demand Manufacturing is approaching. This article tries to follow the trend and use finite element analysis investigating parameters, such as the angle, shift distance, and wire diameter, applied in a CNC bending machine with a swivel plate as the die on the formed cylindrical helical coil springs. The results show that the angle of the swivel plate normal to the wire feeding direction and shift distance have a significant effect on the spring diameter, while the wire diameter has a significant effect on the spring pitch. The result will be used as a reference for the industry to develop On-Demand Manufacturing machinery applications in the future.

1. Introduction

Spring is an important machine element used to connect other machine elements and to provide them movement along with traction, thrust, or torque by its elastic extension, compression, or torsion incorporated to the dynamically or statically load applied between them. Among the sorts and types of spring used in a machine, helical coil springs have been detailed in tables listed in the machinery handbook [1], from which one can easily pick up according to the assigned working parameters. Furthermore, one can readily design his own helical springs as well according to the mechanical design textbook [2]. This is especially meaningful for the customized market era of Industry 4.0 [3], which has been leading to an On-Demand Manufacturing [4]. To response the potential needs, a CNC wire bending machine [5] has been presenting in the market to provide flexible manufacturing a spring made from a wire coil. Even there is CNC program already built in the machine to create the bending path for manufacturing springs, the instinct springback characteristics of the metal wire after bending is, however, not taken into account in the bending path, so that the made spring geometry might not be the same as desired or designed. If an empirical formula or equation can be specifically provided to compensate the springback after the wire bending, it might help the industry to improve their CNC program manufacturing a perfect spring as desired. This article is thus aimed to investigate the influence of the parameters during wire bending to the final geometry of the formed coil spring. Based on the results provided by this paper, a further study can realize the springback compensation during wire bending for manufacturing customized springs.

2. Setup of experiment and simulation

Using a CNC wire bending machine to form a wire into a helical coil spring, the machine feeds the wire from its mandrel, while a swivel plate in front of the mandrel that can be freely shifted and rotated builds a bending moment between the contact point to the wire and the mandrel tip, so that the portion of the wire, where the bending moment engages, bends to form the needed curvature. The swivel plate furthermore exerts a thrust to the wire to push the wire forming the needed pitch. Figure 1 schematically illustrates the principle to manufacture a helical coil spring with a CNC wire bending machine.

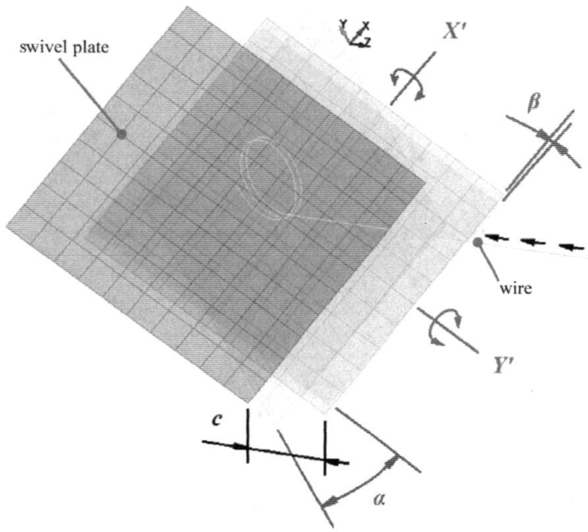

Figure 1. Illustration of the manufacturing process of helical springs by feeding wire and rotating the swivel plate of a CNC wire bending machine.

A wire is fed in Z-direction, while the center of the swivel plate has a given distance c to the feeding mandrel tip, which has the position of the origin at (0, 0, 0). The swivel plate has further its own two independent rotation axes, X' and Y', on its plane. The swivel plate is first located on the original XY-plane, the plane nearest to the triad shown in Figure 1, then rotates first around its X'-axis with an angle α and followed by rotating around its Y'-axis with an angle β, and thereafter is shifted with a distance c in the Z-direction. The declination of the swivel plate is thus defined by the Euler angles to its local Y'-axis, α, and followed by β to its local X'-axis. The normal vector $\vec{n}$ of the swivel plate can be uniquely determined after two times rotation by using matrix manipulation with element rotations as equation (1):

$$\vec{n}=\begin{bmatrix}\cos\beta & 0 & \sin\beta\\ 0 & 1 & 0\\ -\sin\beta & 0 & \cos\beta\end{bmatrix}\begin{bmatrix}1 & 0 & 0\\ 0 & \cos\alpha & -\sin\alpha\\ 0 & \sin\alpha & \cos\alpha\end{bmatrix}\begin{bmatrix}0\\ 0\\ 1\end{bmatrix} \tag{1}$$

Or equation (2) after further manipulations:

$$\vec{n}=\begin{bmatrix}\cos\alpha\sin\beta\\ -\sin\alpha\\ \cos\alpha\cos\beta\end{bmatrix} \tag{2}$$

The setting parameters of this study for the swivel plate are thus defined by only two Euler angles α and β for rotation and one shift distance c for translation.

The commercial finite element software LS-DYNA is applied in this study, which can provide a rigid wall that can efficiently model the swivel plate. The contact between the wire surface to the swivel plate is set as Coulomb friction with a coefficient of 0.125 for usual cold metal forming process, while the contact of the wire surface to itself can be modelled as well. A 65Mn steel having the flow stress σ=400.7$\varepsilon^{0.1428}$ to its equivalent strain ε according to [6] is used to model the wire material. The wire is

200 mm long modelled with the Hughes-Liu cross section integration beam formulation [7] having 16 integration points. To determine the features of a helical coil spring, such as pitch and diameter, after forming, the method used in the industry for determining cylindricity of a shaft is applied [8]. The axis is accordingly calculated and further estimated. Besides the average diameter can be calculated, the average pitch can then be calculated from the individual estimated pitch between two adjacent nodes, in that the estimated pitch is related to their project distance on the spring axis and their sector angle on the π-plane perpendicular to the spring axis. In general, the Hughes-Liu beam effectively generates a constant moment along its length. Thus, meshes need to be reasonably fine to achieve adequate accuracy [9]. However, a finer mesh needs more computation resource. It should be taken a balance between accuracy and efficiency. After verification with several element sizes between 0.05 mm and 1 mm, the element size 0.2 mm is taken for this study. Figure 2 shows the top view and the side view of the formed spring with element size 0.2 mm from its axis.

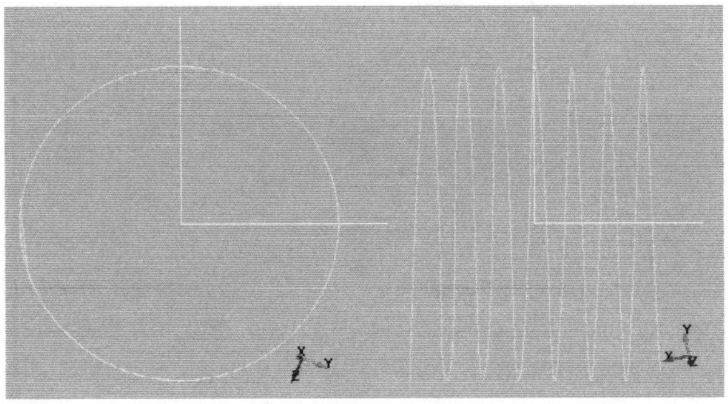

Figure 2. Top view and side view of the formed spring from its axis.

For finite element model with smaller size, the time step, which is proportional to the element size, is smaller as well. To reduce the solving process time, LS-DYNA allows user to set a larger time step by adding element mass, "mass scaling." The time step determined by LS-DYNA for the model with element size of 0.2 mm is 3.42×10^{-8} s. Under this time step, an average diameter in 9.064 mm (standard deviation 0.028 mm) and pitch in 1.018 mm (standard deviation 0.345 mm) are achieved. Compared to the result from the same element size, an average diameter in 9.062mm (standard deviation 0.030 mm) and pitch in 1.018 mm (standard deviation 0.318 mm), which is obtained under a time step of 1.08×10^{-7} s, no significant difference can be noticed, but the computer time is 3.16 times shorter by mass scaling. Even the initial and the following mass scaling might cause more kinetic energy, a quasi-static process is ensured in this study because the ratio of the kinetic energy to the internal energy induced by mass scaling is as small as 2% after initial stage.

To further reduce the computation time, the feeding speed of the wire is accelerated directly at the process beginning to the final speed of 500 mm/s within the first 3 mm feeding. Because the acceleration of the wire feeding might cause additional dynamic behavior, such as vibration, the contact point might be changed during the forming process and a non-uniform diameter might be formed. To freeze the unintended vibration, an artificial viscosity is thus implied in the model, in that the mass damping parameter is set as suggested by the software, $0.4\pi f$, where f is the lower frequency of the vibration determined from the velocity evolution obtained from the model without an artificial viscosity.

3. Results and Discussion

For a constant angle of the swivel plate normal to the *Z*-direction, the influence of different projection angles of the swivel plate to the *X*-direction on the formed spring is investigated. Table 1 lists the setups of the swivel plates, whose normal has an angle 45° to the *Z*-direction but different projection angle to the *X*-direction, for this investigation and table 2 shows their results on the average pitch and diameter as well as the angle of spring axis to the swivel plate normal. Based on the results obtained by the statistical analysis software IBM SPCC, there is no significant relationship of the results to the projection angle of the swivel plate normal. The standard deviation of the pitch and diameter are 0.007 mm and

IOP Conf. Series: Materials Science and Engineering 1222 (2022) 012003 doi:10.1088/1757-899X/1222/1/012003

0.0128 mm, respectively. Comparing them to the standard industrial tolerance, it can be concluded that there is no significant influence of the projection angle to the pitch or the diameter.

Table 1. Euler angles of the swivel plate having an angle 45° to the *Z*-direction but different projection angle to the *X*-direction

Euler angle	projection angle to the *X*-direction (°)									
(°)	0	10	20	30	40	50	60	70	80	90
α	0	7.05	14.00	20.70	27.03	32.80	37.76	41.64	44.14	45
β	45	44.56	43.22	40.89	37.45	32.73	26.57	18.88	9.85	0

Table 2. Pitch, diameter, and axis angle obtained with the swivel plate having an angle 45° to the *Z*-direction but different projection angle to the *X*-direction

	projection angle to the *X*-direction (°)									
	0	10	20	30	40	50	60	70	80	90
pitch (mm)	1.018	1.015	1.025	1.012	1.000	1.010	1.018	1.008	1.022	1.019
diameter (mm)	8.946	8.949	8.932	8.960	8.945	8.960	8.944	8.977	8.945	8.939
axis angle (°)	86.09	84.90	86.17	83.50	86.27	83.28	86.37	84.26	86.23	86.32

On the other hand, for a constant angle of the swivel plate normal projection angles to the *X*-direction, here 45°, the influence of angles of the swivel plate normal to the *Z*-direction on the formed spring is investigated. Table 3 lists the Euler angles of the swivel plates for this investigation and table 4 shows their results on the average pitch and diameter as well as the angle of spring axis to the swivel plate normal. It can be found further that there is a significant relationship of the diameter and the spring axis angle to the normal angle, while there exists no significant influence of the normal angle on the pitch, which can be interpreted as well under that the standard deviation of the pitch is as small as 0.020 mm. It can be observed that the higher the normal angle of the swivel plate to the *Z*-direction, the larger the diameter and the larger the spring axis angle to the swivel plate normal.

Table 3. Euler angles of the swivel plate having a normal projection angle 45° to the *X*-direction but different angle to the *Z*-direction

Euler angle	projection angle to the *X*-direction (°)									
(°)	5	10	20	30	40	50	60	70	80	90
α	3.53	7.05	14.00	20.70	27.03	32.80	37.76	41.64	44.14	-
β	3.54	7.11	14.43	22.21	30.68	40.12	50.77	62.76	76.00	-

Table 4. Pitch, diameter, and axis angle obtained with the swivel plate having a normal projection angle 45° to the *X*-direction but different angle to the *Z*-direction

	projection angle to the *X*-direction (°)									
	5	10	20	30	40	50	60	70	80	90
pitch (mm)	1.009	1.008	1.022	1.016	1.021	1.018	0.990	1.017	0.961	-
diameter (mm)	3.899	4.278	5.021	6.332	7.895	10.17	13.86	21.43	42.52	-
axis angle (°)	83.71	82.12	83.38	83.23	86.07	88.18	89.98	91.23	92.85	-

Table 5 shows the results on the average pitch and diameter as well as the angle of spring axis to the swivel plate normal related to the shift of the feeding mandrel from the swivel plate, *c*. Because smaller shift might cause numerical instability induced by an initial penetration in contact to the swivel plate, 1.5 mm is taken as the smallest shift for a wire diameter of 1 mm in this study, in which the swivel plate has a normal angle 50° to the *Z*-direction and a projection angle 45° to the *X*-direction listed in Table 3 with Euler angles α=32.80° and β=40.12°. It can be found that there is a significant relationship of the diameter to the shift but no significant influence of the shift on the pitch and the spring axis angle. The larger the shift, the larger the spring diameter.

Table 5. Pitch, diameter, and axis angle obtained with different shift of the feeding mandrel to the swivel plate

	shift c (mm)			
	1.5	2	3	5
pitch (mm)	1.020	1.018	1.003	1.039
diameter (mm)	7.407	10.17	15.82	27.02
axis angle (°)	84.88	88.18	85.07	85.71

Table 6 shows the results related to the wire diameter. Due to the fact that smaller shift might cause numerical instability induced by an initial penetration in contact to the swivel plate mentioned in section 3.3, 5 mm is taken as the shift for this study. It can be found that there is a significant relationship of the pitch to the wire diameter, while no significant influence of the wire diameter on the spring diameter and the axis angle.

Table 6. Pitch, diameter, and axis angle obtained with different wire diameter

	wire diameter (mm)			
	0.5	1	2	5
pitch (mm)	1.021	1.039	2.067	5.580
diameter (mm)	27.93	27.02	26.79	27.88
axis angle (°)	86.15	85.71	88.55	76.64

4. Conclusion

A finite element analysis with a commercial code LS-DYNA on forming cylindrical helical coil springs by a CNC wire bending machine with a swivel plate as the forming die has been conducted in this study. Parameters, such as the angle of the swivel plate, shift distance, and wire diameter, on the formed cylindrical helical coil springs. The results show that the angle of the swivel plate normal to the wire feeding direction and shift distance have a significant effect to the spring diameter, while the wire diameter has a significant effect to the spring pitch. The direction of the swivel plate normal projected on the plane perpendicular to the wire feeding direction shows no significant effect on the spring geometries for a certain angle of the swivel plate normal to the feeding direction. As a result, more curvature can be brought into the sheet metal part by using the sawtooth zigzag path than by using the triangle or rectangle zigzag path. The more impression applied to the metal sheet from the bottom roll, the more curvature can be carried out in the part. The smaller the diameter of the impression roll, the more curvature can be achieved as well. However, the thinner and the stiffer the sheet metal, the more curvature can be realized onto the shaped sheet metal part.

References

[1] Oberg E 2020 *Machinery's Handbook Toolbox* (South Norwalk: Industrial Press)
[2] Budynas R and Nisbett K 2020 *Shigley's Mechanical Engineering Design* (New York: McGraw-Hill)
[3] Bauernhansl T, ten Hompel M and Vogel-Heuser B 2014 *Industrie 4.0 in Produktion, Automatisierung und Logistik* (Wiesbaden: Springer)
[4] Daehn G 2021 *Proc. 13th Int'l Conf. on the Technology of Plasticity*
[5] Plussprings Machinery 2017 *ZX-250F* (Tampa: AIM)
[6] Fann K J 2019 *Proc. 6th Taiwan Symp. on Technology of Plasticity* (in Chinese)
[7] Livermore Software Technology Corp. 2019 *LS-DYNA Theory Manual* (Livermore: LSTC)
[8] Lao Y Z, Leong H W, Preparata F P and Singh G 2003 *Precis. Eng.* **27** 4, pp 429-437
[9] Livermore Software Technology Corp. 2013 *LS-DYNA Introductory Examples Manual*, (Livermore: LSTC)

IOP Conf. Series: Materials Science and Engineering 1222 (2022) 012004 doi:10.1088/1757-899X/1222/1/012004

Influence of the construction stages analysis on the distribution of displacement and forces in columns of reinforced columns tall buildings

Raina Boiadjieva

Department of Structural Mechanics
University of Architecture, Civil Engineering and Geodesy
1046 Sofia, Bulgaria

rkirilova_fce@uacg.bg

Abstract. The increasing construction of tall buildings in the world, as well as in Bulgaria, defines the desire for a more precise study of this type of structures. To satisfying this necessity, the possibility of carrying out nonlinear analysis of structures is included in many software packages. The question of the proper definition of material characteristics and the application of computational software in the design of structures with consideration of the stages of construction remains open. In addition to purely research, the problem has also a practical application related to the increasing number of tall buildings at different stages of design and construction. The issue of more accurate modeling of the behavior of tall buildings is discussed, with a focus on the distribution of forces and displacements in columns. A computational model of a tall building is considered. The distribution of forces and displacements in a computational model that accounts for the stages of construction is compared to one in which such consideration is not included. A comparison of the forces and displacements in representative sections of the columns in the two modeling options is presented as a result. Finally, general conclusions and recommendations for the research and design of tall buildings with reinforced concrete structure are given.

1. Introduction

The number of tall buildings in the world is constantly growing. They are the subject of increasing research mainly due to the fact that the effects of vertical and horizontal loads on the structure are much greater than in buildings with the usual number of storeys. The analysis and design are also hampered by the fact that the applicable technical standards are not specifically designed for tall buildings. Defining the main steps and issues in the design of tall buildings taking into account the specific experience of the authors is carried out in [1], [2]. One of the essential features in the design of tall buildings is the necessity to reflect the stages of construction and their influence on the distribution of forces and displacements in the structural elements. Each construction is carried out in separate stages. At each stage of the construction, the structure is deformed under its own weight. The loads from the next phases of the construction are applied on a part of the structure, in which certain displacements, internal forces and stresses have already taken place. Taking the nonlinear behavior of the studied structures into account leads to different in size and sometimes type of deformations. This affects the magnitude and distribution of stresses in the structural elements [4,6].

The nonlinear behavior of a structure is influenced by the technology and speed of execution, the type and geometry of the structure and structural elements, including the location and number of columns and walls in plan, the ratio in the cross-sections between columns and walls, the stiffness of the storey structures, number of storeys, accounting of shrinkage and creep, the properties of the materials and the affect of the environment [3,6].

For tall buildings, the longitudinal deformations and the distribution of normal forces in the vertical structural elements due to gravitational loads have the main affect. As a consequence of the longitudinal deformations a difference in the length of vertical elements is obtained, which leads to redistribution of internal forces in the load-bearing structural elements and the appearance of cracks and deformations.

Traditional for the construction of tall building structures with cores and columns, which are located along the contour of the slabs is considered below. The analysis shows the sensitivity of this type of structures when the construction sequence is taken into account.

2. Description of the structure and the loading

A 42-storey residential building with a total height of 144 meters above lv. 0.000 is presented.

2.1. Description of the structure

The structure consists of reinforced concrete cores and walls, a system of columns and beams along the contour of the floor slabs. The architectural solution requires a change in the contour of each floor slab, which leads to a change in the geometry of part of the walls and columns. The soil conditions are not taken into account and the structures supports are considered as fixed. The walls are 70 cm thick, the dimensions of the columns are 100/100 cm, the beams are 60/70 cm and the slabs are 18 cm thick. The properties of concrete grade C35/45 and reinforcement steel B500 are used in the calculations (Fig.1). In order to clearly highlight the effects of redistribution of internal forces the dimensions of the cross-sections of the columns and walls are assumed to be constant throughout the height of the structure.

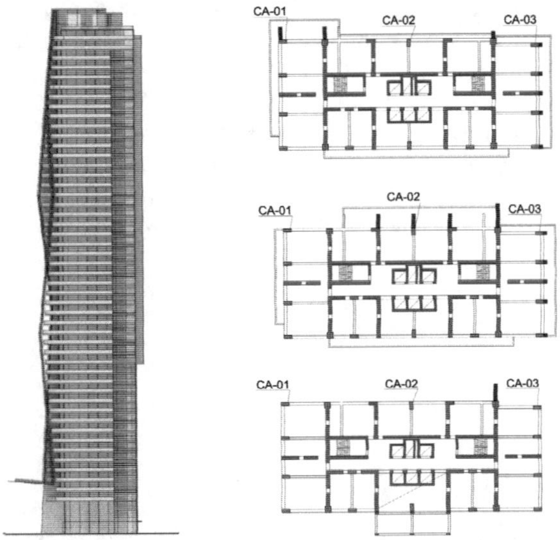

Figure 1. A facade and formworks plans

2.2. Defining load function

The load on the structure is modelled and is divided into two main groups: Fk – self weight of the structural elements; Fдоп - load from non-bearing structures and other long-term loads. Live loads are not included because the structure is studied at the time of its construction.

The load for the i-th floor would be:

$$F_i = (F_k + F_{доп}) * (n - i + 1) \quad (1)$$

The influence of time or speed and sequence of execution can be presented by the rate of application of the loads or the average rate of change of Fi [3].

$$C_i = \frac{F_i}{t_i}, \text{ where} \tag{2}$$

C_i – speed of the construction process;
F_i – service loads (in this case without live load);
t_i – the total time required to complete the structure.
Under these initial conditions, the load can be defined as a time function:

$$F_i(t) = C_{k,i} * t_{k,i} + C_{доп,i} * t_{доп,i} \tag{3}$$

This function takes into account that different construction processes require different speeds and start at different times. I.e. $C_k \neq C_{доп}$ and $t_{0,k} \neq t_{0,доп}$.

The loads are as follows: F_k – is automatically calculated by the software; $F_{доп}$ – 4,50 kN/m^2.

The accepted speed of the construction processes is $C_{k,i} = C_{доп,i} = 1$ floor per 20 days, as $t_{0,k} = 0$ day and $t_{0,k} = 80$-th day.

Only static analysis of the frame is carried out. The coefficient $\psi = 1$ as the results are used only for comparison of the two different ways of loading. The construction stages are considered in the analysis correspond to a possible construction schedule.

3. Numerical model

The analysis software SAP2000 [7] is used for modelling and structural calculations. The beams and the columns are modelled with Frame elements. The slabs and the walls are modelled with Shell elements (Fig.2).

For the purposes of this research, four variants of calculation models have been created and analyzed, in which the number of storeys of the building varies - 10, 20, 30 and 42 floors.

To take into account the embedded reinforcement in the reinforced concrete structure, the modulus of elasticity of the concrete is increased by 5% [5].

Two different variants of defining the loading are considered and modelled numerically:

- The loads are defined and combined in a way that is usual in the engineering. The deformations due to the self-weight of the structural elements during construction are not taken into account and it is also not considered that the rest of the loads are applied to the structure after the elements are deformed due to their self-weight. The results are presented in the following tables under the name "C";

- The second variant the loads $F_{доп}$ are applied to the already constructed structure that is deformed because of the self-weight. The results are presented in the following tables under the name "CS".

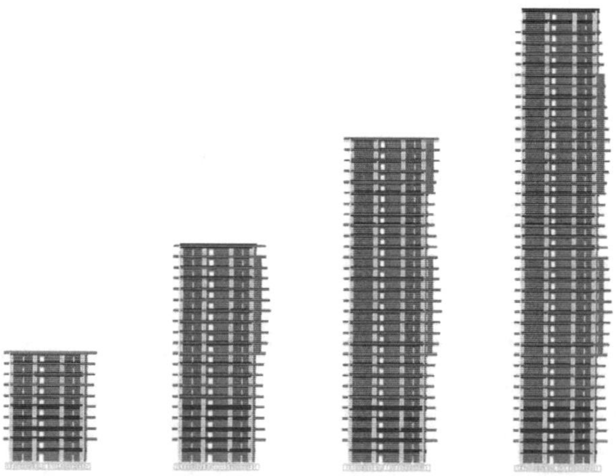

Figure 2. Computational models – versions with different number of storeys

4. Analysis and comparison of the results

The main results of the calculations of the two variants of modelling are compared. The loads and the load cases are the same, but the number of storeys vary for every of the different models. The results present the vertical displacements and the normal forces at 3 of the columns (Fig.1) on specific floors – 1-st, 10-th, 20-th, 30-th and 41-st. A comparative analysis is given of whether and what influence the number of storeys have on the distribution and magnitude of displacements and internal forces for the same loads applied under the same law.

The two sets of results reveal differences in the vertical displacements and the normal forces. Comparison of the differences in % between the results of the two solutions is presented in the Figure 4.

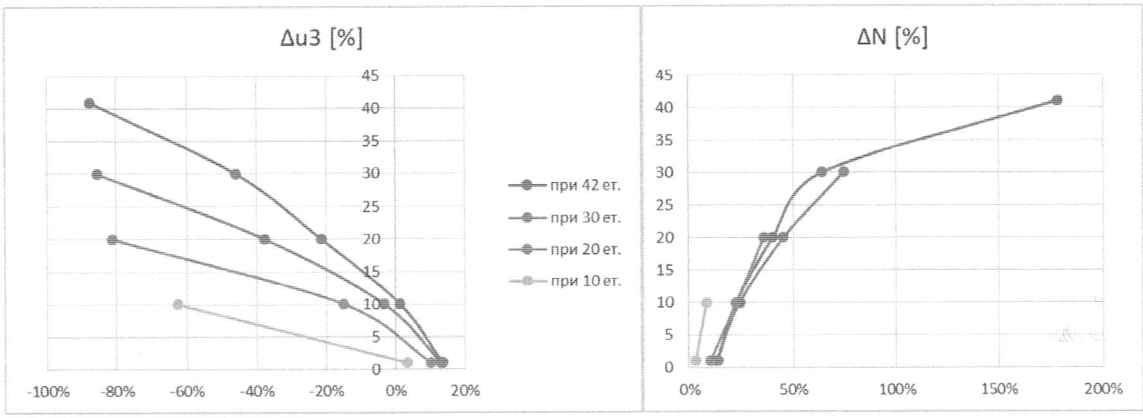

Figure 3. The differences in % between the results of the two solutions

The main conclusions that follow from the presented results are:

- The vertical deformations in the columns are significantly smaller when the construction stages are taken into account;
- Normal forces are significantly greater when the construction stages are taken into account;
- In the case of building structures with up to 10-15 storeys, the consideration of the construction stages in the computational model does not have a significant impact on the distribution of forces in the columns.

In tall buildings, the increased number of floors changes the way how displacements and forces are being distributed compared to structures with fewer storeys.

In the nonlinear structural analysis considering the stages of construction, it is required to ensure the proper definition of the loads and the speed with which they are applied.

References

[1] Traykov A., *Main features and issues in the design of tall buildings in Bulgaria*, DCB2020 XI X I International Scientific Conference Civil Engineering Design And Construction (Theory and Practice for a Sustainable Future), Varna,Bulgaria 10-12 September 2020

[2] Traykov A., Traykova M., *Tall residential buildings – practical implementation of design concepts and code requirements*, 30th Biennial National Conference of the Concrete Institute of Australia (Concrete 2021), to be held virtually, from 5-8 September 2021

[3] Kozak J. *Tall buildings structures*, Stroyizdat, Moscow

[4] Bui Thanh Dat, Al. Traykov, M. Traykova, *Shear lag effect and its effect on the design of high-rise buildings*, High Rise Constructions 2017, September 2017, Samara, Russia, E3S Web of Conferences 33,02001

[5] Manoilov L., *Reinforced concrete*, Tehnika, Sofia, Bulgaria 1991

[6] Lagos, R., Kupfer, M., Sanhueza, S., Cordero, F., *Costanera Center - Shortenings due to elastic deformation, creep and shrinkage of concrete in a 300-m tall building*, ICJ, India, 2012
[7] SAP2000, Version 20.2.0, Computers and Structures Inc, 2019

Acknowledgments
The author would like to express her acknowledgements to Dr. Alexander Traykov head of "AVT Consult" for the opportunity to be part of the design team that worked on the structure of the presented building in period 2015/2017.

IOP Conf. Series: Materials Science and Engineering 1222 (2022) 012005 doi:10.1088/1757-899X/1222/1/012005

Synergistic effect of essential oil distillates on the corrosion protection of GX4CrNiMo16-5-1 stainless in acid-chloride electrolyte

R T Loto[1]

Department of Mechanical Engineering, Covenant University, Ogun state, Nigeria

Email: tolulope.loto@covenantuniversity.edu.ng

Abstract. The inhibition capability of admixed rosemary oil with cinnamon cassia oil (RC), and grapefruit oil with lemongrass oil (GL) on GX4CrNiMo16-5-1 martensitic stainless steel (GX4ST) corrosion was studied in 6 M H_2SO_4/3.5% NaCl electrolyte by potentiodynamic polarization and optical microscopy characterization. Results show both admixed distillates performed effectively on GX4ST. The corrosion rate of GX4ST without the distillates at 2.82 mm/y significantly decreased to average value of 0.04 mm/y and 0.05 mm/y analogous to average inhibition efficiency of 98.64% and 98.23% for RC and GL compound. The oil distillates exhibit consistent anodic corrosion inhibition behavior at all concentrations studied. Corrosion thermodynamic calculations showed the distillates chemisorbed on the martensitic steel with Gibbs free energy values greater than -46.01 $Kjmol^{-1}$ with respect to Langmuir isotherm equation which indicates negligible of lateral interaction effect among inhibitor molecules and surface coverage protection mechanism. Optical image representations of GX4ST after corrosion without the distillates showed severe surface degradation of general and localized corrosion. This contrast the images of the inhibited steel surface which were closely similar to the steel image before corrosion.

1. Introduction

The consequences of corrosion of metallic alloys is enormous [1]. The most widely used metallic alloys are ferrous based especially stainless steels. Significant number of manufactured products consists of stainless steel. Stainless steels are recyclable, exhibits excellent mechanical properties, interfacial equilibrium, and thermal deformation resistance [2-4]. The steels are extensively utilized for structural application in nuclear, chemical processing, desalination, fertilizer production and petrochemical industries by reason of their extraordinary resilience against corrosion compared to carbon steels [5, 6]. The resilience of stainless steels against corrosion is due to the evolution of an insulating and inert oxide film on the steel. The oxide results from the presence of alloying elements within the steel metallurgical structure especially Cr and Ni which chemically combines with dissolved O_2 to passivate the steel surface. Electrochemical reaction of stainless steels with anions such as SO_4^{2-}, Cl^-, $S_2O_3^{2-}$, NO^-_3 etc. reacts with the outer electrons on the steel's exterior destabilizing the protective film especial at regions with flaws and discontinuities. Circumstances resulting in the disintegration of the protective oxide on stainless steels are prevalent in astringent industrial conditions couples with extreme temperature and high velocity fluid movement. Sequel to the collapse of passive films localized corrosion degradation majorly responsible for failure of stainless steels initiates and progresses [7, 8]. Pitting corrosion occurs due to the evolution and increase in microscopic holes on stainless steel exterior which is significantly destructive to the operational lifespan of the steel in service especially in aqueous environments [9, 10]. The existence of SO_4^{2-} and Cl^- anions in aqueous environments causes instability of the protective oxide on steels leading to selective deterioration i.e. pitting corrosion [11]. The electrochemical performance of stainless steels analogous to the evolution, resilience and disintegration of their passive film is fundamental to the comprehension of the pitting corrosion resilience of these alloys. Mitigation against pitting corrosion is affiliated with cost consequences due to repair of metallic structures and components, and sustainability of their lifespan and reliability. Pitting corrosion is acute, insidious and undetectable which increases the difficulty to predict and design against [12]. Research on pitting corrosion evolution and progression on stainless steels have provided insight to the present understanding of its phenomena and have contributed to appropriate material selection for industrial application [13-18]. Yi et al [19] evaluated the pitting resistance of type 316 steel in specific chloride solutions and established that the critical pitting potential of the steel substantially depends on induced electrochemical factors. Hodge and Wilde [20] ascertained that Ni-Cr alloys (10% Cr content) exhibited passivation coupled with adequate resilience against pitting in sulphate-chloride solution. Piron et al [21] also ascertained correlating behavior between Ni 200 and Inconel 600 steel in sulphate solution. Friend [22] studied the disintegration of the insulating film on alloy 22 at critical chloride concentration. Investigation on metastable and stable pit propagation coupled with the influence of electrochemical factors have seen considerable progress with several mechanisms proposed to explain collapse of the passive film [23-27]. The resilience of these alloys against the damaging effect anionic species largely depends on the attributes and composition of their insulating film, repassivation tendency, metastable pitting behaviour and resilience to pitting corrosion. The ability of the protective oxide on stainless steels to with stand collapse and/or repassivate after breakdown in aqueous corrosive environments strongly influences the resilience of the steels to pitting corrosion. This can be significantly improved upon through the utilization of chemical fluids identified as corrosion inhibitors. Corrosion inhibitors substantially decrease the corrosion rate of metallic alloys in corrosive fluids [28, 29]. Inhibitors exhibit definite reaction behaviour analogous to specific alloys, industrial condition, concentration range etc. which influences their effectiveness. Traditional corrosion inhibitors (e.g. chromates, nitrates, nitrites, arsenic, phosphates etc.) are toxic [30-32]. Compounds of organic origin are proven alternatives but expensive and slightly toxic [33-39]. The relevance of sustainable, safe and cost-effective compounds cannot be overemphasized. Plants distillates have shown in previous research to be promising [40-42]. However, their inhibition performance against localized corrosion reaction phenomena needs to be ascertained. This manuscript focusses on the effect of plant distillates on the pitting corrosion evaluation and inhibition of GX4CrNiMo16-5-1 martensitic stainless steel.

2 Experimental methods

GX4CrNiMo16-5-1 martensitic stainless-steel plate (GX4ST) was acquired from Vienna University of Technology, Austria with average nominal (wt.%) content laid out in Table 1. The steel plate was cut to 12 test pieces with average dimension of 1 cm^3. They were subsequently smoothened with emery papers with 120, 240, 320, 600, 800, 1200 and 2500 grit before being washed with deionized H_2O and acetone. Grapefruit and lemon oil distillates were combined in ratio 1:1 and concocted in cubic concentrates of 1%, 2%, 3%, 4% and 5% in 200ml of 6M H_2SO_4/3.5% NaCl electrolyte while rosemary and cinnamon cassia were combined in ratio 1:1 and similarly prepared.

Table 1. Nominal (wt. %) content of GX4ST

Element	C	Si	Mn	P	S	Cr	Mo	Ni	Ti	N	Cu	V	Fe
Content	0.06	0.8	1	0.04	0	17	1.5	6	0	0	0.3	0.1	73.21

Potentiodynamic polarization investigations were executed with a triple electrode cable design inside a beaker filled with 200 mL of the electrolyte and joined to Digi-Ivy 2311 potentiostatic device. GX4ST electrodes encased in solidified resin with visible surface area of 1 cm^2. Polarization plots were observed at sweep rate of 0.0015V/s from set potential of −0.5V and +1.5V. Pt wire was engaged as the counter electrode and Ag/AgCl as the standard electrode. Corrosion current density, J_c (A/cm^2) and corrosion potential, E_c (V) data were gotten from Tafel estimation approach. The corrosion rate C_R, (mm/y) and the inhibition efficiency, η (%) were computed from the numerical formula;

$$C_R = \frac{0.00327 \times J_c \times E_q}{D} \quad (1)$$

D represents density (g/cm^3) E_q represents equivalent of GX4ST. 0.00327 represents corrosion rate constant for computations in mm/y. Optical microscopy analysis was done with Omax trinocular metallurgical microscope for morphological analysis of the steel surface.

3 Results and discussion

3.1 Potentiodynamic polarization studies

Potentiodynamic polarization plots of GX4ST corrosion in 6M H_2SO_4/3.5% NaCl solution at 1-5% RC and GL concentration are laid out in Fig. 1(a) and (b). Table 2 exhibits the polarization data obtained. The corrosion rate of GX4ST at 0% RC and GL concentration (2.82 mm/y relating to corrosion current density of 2.62 x $10^{-4}$$A/cm^2$) significantly differs from the values obtained at 1-5% concentration of the inhibitors preceding from oxidation of the steel surface by SO_4^{2-} and Cl^- anions. The anions are responsible for the release of Fe^{2+} ions into the acid-chloride electrolyte coupled with Cr depletion on the steel's protective film. Consequentially, the chemical fusion of H_2 atoms leads to evolution of H_2 gas and O_2 reduction reactions which are also responsible for selective degradation of the steel. Observation of GX4ST polarization plots at 0% inhibitor concentration in Fig. 1(a) and (b) shows the anodic-cathodic polarization slopes are substantially greater than the slopes at other inhibitor concentration due to action redox reaction mechanism during potential scanning. At 1% RC concentration in Fig. 1(a), significant decrease in the anodic-cathodic plot configuration are visible due to interference and suppression of the redox reaction mechanism. This observation coincides with corrosion rate of 0.03 mm/y and corrosion current density of 3.04 x 10^{-6} A/cm^2. RC concentration had negligible effect on its inhibition performance with inhibition efficiency result remaining generally similar. This proves the inhibition performance of RC is concentration dependent. This claim is proven from the similarities in configuration of the anodic polarization plots at 1-5% RC concentration and the cathodic plots at 2-5% RC concentration. It signifies the inhibition performance of RC is under activation control due to stability of the inhibition action and behaviour of RC analogous to concentration. Corrosion rate of GX4ST varies between 0.03 mm/y and 0.04 mm/y while the inhibition efficiency varies between 98.84% and 98.53%. The corrosion potential of the polarization plots at specific RC concentration tends to electropositive values due to dominant anodic inhibition behaviour. However, the optimal difference between the corrosion potential of GX4ST polarization plot at 0% RC and specific RC concentration is 103 mV signifying RC exhibits anodic inhibition behaviour. This shows that protection of the steel by protonated RC molecules occurred through adsorption. This is further confirmed from the pseudo passivation behaviour of the anodic polarization plots.

The performance of GL inhibitor is generally similar to RC with respect to corrosion rate and inhibition efficiency values, which varied between values of 0.06 mm/y – 0.04 mm/y and 97.94% - 98.71%. However, variation of the anodic-cathodic plots with respect to GL concentration shows GL inhibition behaviour steers significantly away from activation control mechanism. The slopes of the plot configuration show the mechanism of GL inhibition is significantly influenced by the redox reactions taking place on the steel surface. The deviation in corrosion potential to electropositive values shows GL inhibits GX4ST surface through surface coverage with optimal difference in corrosion potential between GL inhibited steel and the steel at 0% GL concentration is 90 mV which shows GL is also an anodic type inhibitor. Comparing the performance of the inhibitor compounds it's clearly visible that the performance of the inhibitors is generally similar and highly effective.

Table 2. Potentiodynamic polarization output of GX4ST corrosion in 6 M H_2SO_4/3.5% NaCl solution at 1-5% RC and GL concentration

RC Inhibitor									
Sample	**RC Conc. (%)**	**LCS C_R (mm/y)**	**RC ξ_F (%)**	**C_I (A)**	**C_J (A/cm²)**	**C_P (V)**	**R_p (Ω)**	**B_c (V/dec)**	**B_a (V/dec)**
A	0	2.82	0	2.96E-04	2.62E-04	-0.364	86.73	-9.115	12.490
B	0.5	0.03	98.84	3.44E-06	3.04E-06	-0.261	7479.00	-8.635	13.370
C	1	0.04	98.53	4.36E-06	3.86E-06	-0.328	5894.00	-7.056	9.081
D	1.5	0.03	98.91	3.23E-06	2.86E-06	-0.310	7961.00	-5.662	9.139
E	2	0.04	98.41	4.71E-06	4.17E-06	-0.285	5450.70	-7.698	13.167
F	2.5	0.04	98.53	4.37E-06	3.86E-06	-0.285	5886.00	-7.149	11.300
GL Inhibitor									
Sample	**RCC Conc. (%)**	**LCS C_R (mm/y)**	**RCC ξ_F (%)**	**C_I (A)**	**C_J (A/cm²)**	**C_P (V)**	**R_p (Ω)**	**B_c (V/dec)**	**B_a (V/dec)**
A	0	2.82	0	2.96E-04	2.62E-04	-0.364	86.73	-9.115	12.490
B	0.5	0.06	97.94	5.74E-06	5.08E-06	-0.274	4477.00	-10.660	28.710
C	1	0.05	98.06	5.42E-06	4.79E-06	-0.282	4815.00	-11.550	14.290
D	1.5	0.06	97.99	5.63E-06	4.98E-06	-0.324	5387.80	-9.321	12.600
E	2	0.04	98.43	4.38E-06	3.88E-06	-0.342	6078.00	-10.290	14.000
F	2.5	0.04	98.71	3.59E-06	3.18E-06	-0.274	7158.00	-9.496	25.270

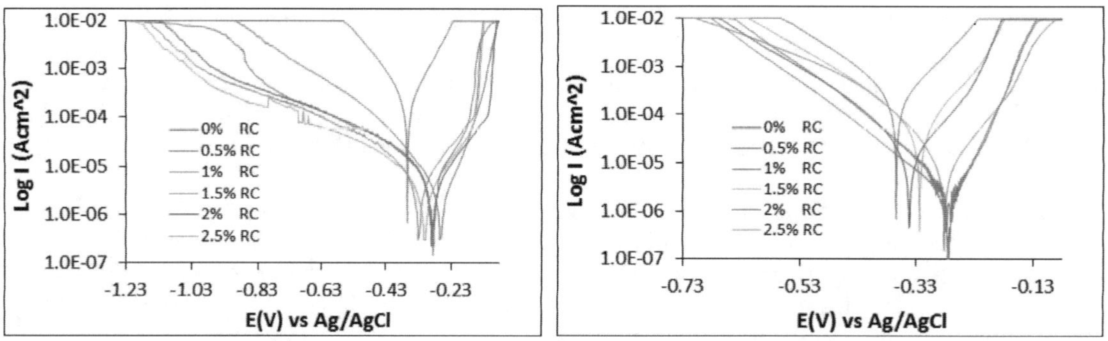

(a) (b)

Figure. 1 Potentiodynamic polarization plots of GX4ST corrosion in 6 M H2SO4/3.5% NaCl solution at 1-5% (a) RC concentration and (b) GL concentration

3.2 Corrosion thermodynamic studies

Adsorption isotherms provide insight into the mechanism of inhibitor-metal interaction leading to stifling of the redox reactions identified with corrosion. Inhibitor protection of metallic alloys occurs through electrostatic attraction between protonated inhibitor molecules in the electrolyte solution and valence electrons on the ionized steel surface. Strong electrostatic attraction culminates in effective corrosion inhibition. However, this assertion is subject to the ionic behaviour of corrosive species as most organic corrosion inhibitors depends on preadsorption of the corrosive species onto the steel before electrostatic attraction of the inhibitor onto the steel surface. Secondly, the organo-metallic complexation is also an important factor i.e. covalent bonding between the RC/GL molecules and the ionized steel surface stifles the corrosion reaction mechanisms. This reaction mechanism is termed chemisorption. In cases where the bonding is weak, it falls under Van der Waals interaction termed physisorption [43]. Data obtained from potentiodynamic polarization was evaluated with conventional adsorption isotherms. Only Langmuir isotherm models prove to be most applicable with interaction coefficients greater than 0.9 for RC and GL inhibitors. The interaction coefficient values are laid out in Table 3 below;

Table 3. Analogy of interaction coefficient values for Langmuir isotherm in 6 M H_2SO_4/3.5% NaCl for RC and GL inhibitor

Adsorption Isotherm	RC Inhibitor	GL Inhibitor
Langmuir	1	0.9999

The Langmuir isotherm plots obtained from the fluid electrolyte in the presence of the inhibitors are presented in Figs. 2(a) and (b). Langmuir isotherm indicates that adsorption of inhibitor molecules occurs at the same definite reaction sites in the absence lateral interaction among inhibitor molecules with respect to mono layer adsorption according to the equation below;

$$\theta = \left[\frac{K_{ads}C_{Inh}}{1+K_{ads}C_{Inh}}\right] \quad (2)$$

K_{ads} represents equilibrium constant of adsorption and C_{Inh} represents molar concentration of inhibitor compound [44].

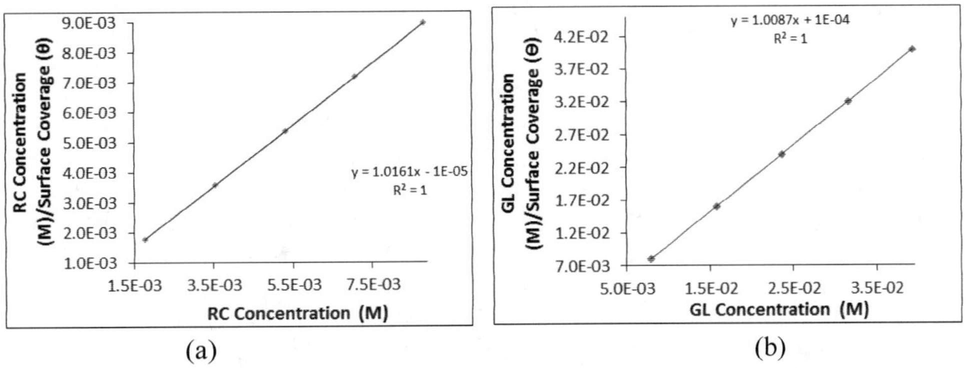

(a) (b)

Figure. 2 Langmuir adsorption isotherm plots from 6 M H_2SO_4/3.5% NaCl solution in the presence of (a) RC inhibitor and (b) GL inhibitor

Data output of Gibbs free energy (ΔG) calculated from equation 5 are displayed in Table 4. The data indicates the nature, potency and stability of adsorption mechanism of RC and GL inhibitor on GX4ST in the acid-chloride solution. The equilibrium constant of adsorption (K_{ads}) was computed from Langmuir isotherm equation.

$$\Delta G_{ads} = -\,2.303RT \log\,[55.5K_{ads}] \quad (3)$$

55.5 indicate molar concentration of H_2O in the acid solution, R indicates universal gas constant and T indicates absolute temperature. The ΔG data from the acid solutions in the presence of RC and GL inhibitors shows the inhibitor interaction mechanism occur through chemisorption adsorption [45]. The ΔG values associated with RC inhibition action are slightly higher than the corresponding values for GL inhibitor which signifying higher adsorption strength. The general slight decrease in value in the presence of both inhibitors results from lateral repulsion effect between protonated inhibitor molecules which in effect is negligible.

Table 4. Data output for Gibbs free energy (ΔG) and equilibrium constant of adsorption (K_{ads}) for RC and GL adsorption on GX4ST in 6 M H_2SO_4/3.5% NaCl

			RC Inhibitor			**GL Inhibitor**		
GX4ST Samples	**Inhibitor Conc. (%)**	**Inhibitor Conc. (M)**	**Surface Coverage (θ)**	**Equilibrium Constant of adsorption (K)**	**Gibbs Free Energy, ΔG (Kjmol^{-1})**	**Surface Coverage (θ)**	**Equilibrium Constant of adsorption (K)**	**Gibbs Free Energy, ΔG (Kjmol^{-1})**
A	0	0	0	0	0	0	0	0
B	0.5	1.77E-03	0.988	48278217.5	-53.79	0.971	4291032.0	-47.80
C	1	3.53E-03	0.985	18967843.5	-51.48	0.990	6149334.6	-48.69
D	1.5	5.30E-03	0.989	17147218.5	-51.23	0.988	3468913.5	-47.27
E	2	7.07E-03	0.984	8770399.9	-49.57	0.989	2770338.8	-46.71
F	2.5	8.84E-03	0.985	7576552.8	-49.21	0.988	2087012.9	-46.01

3.3 Optical image analysis

Optical images of GX4ST prior to and after corrosion assessment are exhibited from Fig. 3(a) to 5(b). Fig. 3(a) presents the image after metallographic preparation prior to corrosion test. Fig. 3(b) shows the morphology of GX4ST after corrosion in 6M H_2SO_4/3.5% NaCl. Extensive surface degradation is clearly visible due to oxidation of the steel exterior by the corrosive action of SO_4^{2-} and Cl^- anions. Corrosion pits are also visible due to localized electrochemical deterioration at definite areas on the steel where the passive protective film weakened. Fig. 4(a) to 5(b) exhibits the morphology of GX4ST after corrosion in the presence of 1% and 5% RC and GL concentrations respectively. The morphologies show effective inhibitor protection as they contrast the degraded morphology in Fig. 3(b). The corrosive anions where hindered from diffusing unto and reacting with the steel surface. However, miniature surface degradation and corrosion pits are still partially visible due to preadsorbed corrosive anions which aided the electrostatic attraction of the protonated inhibitor molecules. However, the surface morphology of the inhibited steels is well protected and assents with the results from potentiodynamic polarization.

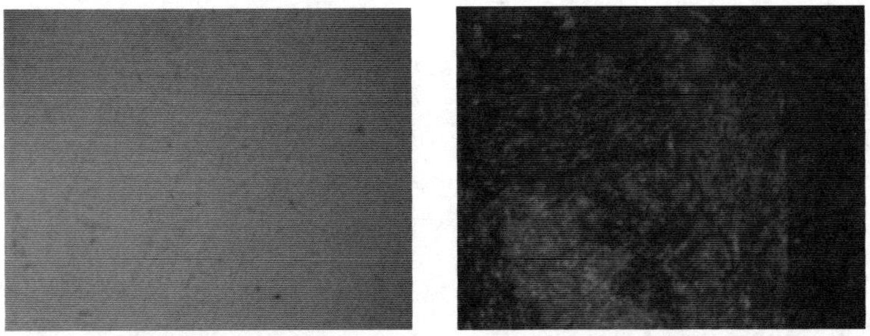

(a) (b)

Figure. 3 Optical images of GX4ST (a) prior to corrosion analysis and (b) after corrosion from 6 M H_2SO_4/3.5% NaCl solution

(a) (b)

Figure. 4 Optical images of GX4ST after corrosion from 6 M H_2SO_4/3.5% NaCl solution at (a) 1% RC concentration and (b) 1% GL concentration

(a) (b)

Figure. 5 Optical images of GX4ST after corrosion from 6 M H_2SO_4/3.5% NaCl solution at (a) 6% RC concentration and (b) 6% GL concentration

4. Conclusion

Admixture of rosemary oil with cinnamon cassia oil, and grapefruit oil with lemongrass oil effectively inhibited the corrosion of GX4CrNiMo16-5-1 martensitic stainless steel in chloride-sulphate solution. Both admixed oil distillates performed adequately at specific concentrations with inhibition efficiency values generally signifying stability of the inhibitive electrochemical behavior of the oil distillates in the electrolyte. The admixed oil distillates exhibited anodic type inhibition effect resulting in surface coverage over the entire steel. The oil distillates adsorbed onto the steel by chemisorption reaction process through covalent bonding and strong electrostatic attraction. Optical images of the inhibited steel significantly differ from the non-inhibited and corroded steel due to absence of protonated oil distillates to counterbalance the combined destructive effect of chlorides and sulphates in the electrolyte.

Acknowledgement

The author recognizes the support of Covenant University Ota, Ogun State, Nigeria for this research.

References

[1] Palaniswamy BRN, Rengaswamy NS and Jayachandran M (2005) A review of differing approaches used to estimate the cost of corrosion (and their relevance in the development of modern corrosion prevention and control strategies) *Anti Corros. Method M* **52(1)** 29-41.
[2] Reuter MA, Heiskanen K, Boin U, Van Schaik A, Verhoef E, Yang Y and Georgalli G (2005) The metrics of material and metal ecology, Harmonizing the resource, technology and environmental cycles, Elsevier Science Vol. 16.
[3] Christopher R (2011) Sustainability and value of steel recycling in Uganda *J. Civ. Eng. Constr. Technol.* **2(10)** (2011).
[4] Tan Q and Li J (2015) Recycling metals from wastes: A novel application of mechanochemistry *Environ. Sci. Technol.* **49(10)** 5849–5861.
[5] Caron P and Khan T (1999) Evolution of Ni-Base Superalloys for single crystal gas turbine blade applications *Aerosp. Sci. Technol.* **3** 513-523.
[6] Hashizume R, Yoshinari A, Kiyono T, Murata Y and Morinaga M (2007) Development of Ni-based single crystal superalloys for power generation gas turbines *Energ. Mater. Mater. Sci. Eng. Energ. Syst.* **2(1)** 5-12.
[7] McCafferty E (2003) Sequence of steps in the pitting of aluminum by chloride ions *Corros. Sci.* **45** 1421-1438.
[8] Abdel Rehim SS, Hassan HH and Amin MA (2004) Chronoamperometric studies of pitting corrosion of Al and (Al–Si) alloys by halide ions in neutral sulphate solutions *Corros. Sci.* **46** 1921-1938.
[9] Loto RT and Loto CA (2017) Corrosion behaviour of S43035 ferritic stainless steel in hot sulphate/chloride solution *J. Mater. Sci. Technol.* doi.org/10.1016/j.jmrt.2017.07.004.
[10] Li S, Xu Y, Li H and Guan X (2014) Uniform and pitting corrosion modeling for high-strength bridge wires *J. Bridge Eng.* **19(7)**. doi.org/10.1061/(ASCE)BE.1943-5592.0000598
[11] Nieuwoudt MK, Comins JD and Cukrowski I (2012) Analysis of the composition of the passive film on iron under pitting conditions in 0.05 M NaOH/NaCl using Raman microscopy in situ with anodic polarisation and MCR-ALS *J. Raman Spectrosc.* **43(7)** (2012), p. 928-938. doi.org/10.1002/jrs.3109.
[12] Guo P, La Plante EC, Wang B, Chen X, Balonis M, Bauchy M and G. Sant G (2018) Direct observation of pitting corrosion evolutions on carbon steel surfaces at the nano-to-micro- scales *Sci. Rep.* 8(7990). doi.org/10.1038/s41598-018-26340-5
[13] Loto RT and Loto CA (2017) Potentiodynamic polarization behavior and pitting corrosion analysis of 2101 duplex and 301 austenitic stainless steel in sulfuric acid concentrations *J. Fail. Anal. Prev.* **17(4)** 672–679.
[14] Loto RT and Loto CA (2015) Pitting corrosion inhibition of type 304 austenitic stainless steel by 2 amino 5 ethyl 1,3,4 thiadiazole in dilute sulphuric acid Prot Met Phys Chem+ 51(4) (2015), p. 693–700.
[15] Loto RT (2013) Pitting corrosion evaluation of austenitic stainless steel type 304 in acid chloride media *J. Mats. & Environ. Sci.* **4(4)** 448-459.
[16] Tian W, Du N, Li S, Chen S and Wu Q (2013) Metastable pitting corrosion of 304 stainless steel in 3.5% NaCl solution *Corros. Sci.* **85** 372-379.
[17] Yi Y, Cho P, Al Zaabi A, Addad Y and Jang C (2013) Potentiodynamic polarization behaviour of AISI type 316 stainless steel in NaCl solution *Corros. Sci.* **74** 92–97.
[18] Hodge FG and Wilde BE (1970) Effect of chloride ion on the anodic dissolution kinetics of chromium–nickel binary alloys in dilute sulfuric acid *NACE Corrosion* **26(6)** 246-250.
[19] Piron DL, Koutsoukos EP and Nobe K (1969) Corrosion behavior of nickel and inconel in acidic chloride solutions *NACE Corrosion* **25(4)** 151-156.
[20] Friend WZ (1980) Nickel-Chromium-Molybdenum Alloys, Wiley-Interscience, New York.

[21] Pahlavan S, Moazen S, Taji I, Saffar K, Hamrah M, Moayed MH, Mollazadeh S and Beidokhti S (2016) Pitting corrosion of martensitic stainless steel in halide bearing solutions *Corros. Sci.* **112** 233–240.
[22] Lu K and Lu J (2004) Nanostructured surface layer on metallic materials induced by surface mechanical attrition treatment *Mater. Sci. Eng. A.* **375-377** 38-45.
[23] Pistorius PC and Burstein GT (1992) Metastable Pitting Corrosion of Stainless Steel and the Transition to Stability *Philos. Trans.: Phys. Sci. Eng.* **341** 531–559.
[24] Pistorius PC and Burstein GT (1992) Growth of corrosion pits on stainless steel in chloride solution containing dilute sulphate *Corros. Sci.* **33** 1885–1897.
[25] Ernst P, Laycock NJ, Moayed MH, Newman RC (1997) The mechanism of lacy cover formation in pitting *Corros. Sci.* **39** 1133–1136.
[26] Laycock NJ and Newman RC (1998) Temperature dependence of pitting potentials for austenitic stainless steels above their critical pitting temperature *Corros. Sci.* **40** 887–902.
[27] Soltis J (2015) Passivity breakdown, pit initiation and propagation of pits in metallic materials – Review *Corros. Sci.* **90** 5–22.
[28] Popov BN (2015) Chapter 14 - Corrosion Inhibitors, Corrosion Engineering Principles and Solved Problems, Elsevier Inc., Amsterdam, pp. 581-597. https://doi.org/10.1016/B978-0-444-62722-3.00014-8.
[29] Palanisamy G (2018) Corrosion Inhibitors, IntechOpen. http://doi.org/10.5772/intechopen.80542.
[30] Sastri VS (2014) Types of corrosion inhibitor for managing corrosion in underground pipelines, Underground Pipeline Corrosion, Elsevier Inc., Amsterdam, pp. 166-211. https://doi.org/10.1533/9780857099266.1.166.
[31] Chen Y and Yang W (2019) Formulation of Corrosion Inhibitors. http://doi.org/ 10.5772/intechopen.88533.
[32] Dariva CG and Galio AF (2013) Corrosion Inhibitors – Principles, Mechanisms and Applications, IntechOpen. http://doi.org/10.5772/57255.
[33] Ahmed SK, Ali WB and Khadom AA (2019) Synthesis and investigations of heterocyclic compounds as corrosion inhibitors for mild steel in hydrochloric acid *Int. J. Ind. Chem.* **10** 159–173.
[34] Singh A, Talha M, Xu X, Sun Z and Lin Y (2017) Heterocyclic corrosion inhibitors for J55 steel in a sweet corrosive medium *ACS Omega* **2(11)** 8177-8186.
[35] Singh A, Ansari KR, Quraishi MA, Kaya S and Banerjee P (2019) The effect of an N-heterocyclic compound on corrosion inhibition of J55 steel in sweet corrosive medium *New J. Chem.* **43** 6303-6313.
[36] Sundaram RG and Sundaravadivelu M (2016) Anticorrosion Activity of 8-Quinoline Sulphonyl Chloride on Mild Steel in 1 M HCl Solution *Journal of Metallurgy* 8095206. https://doi.org/10.1155/2016/8095206.
[37] Loto RT (2017) Study of the synergistic effect of 2-methoxy-4-formylphenol and sodium molybdenum oxide on the corrosion inhibition of 3CR12 ferritic steel in dilute sulphuric acid *Results in Phys.* **7** 769-776.
[38] Loto CA and Loto RT (2013) Effect of dextrin and thiourea additives on the zinc electroplated mild steel in acid chloride solution *Int. J. Elect. Sci.* **8(12)** 12434-12450.
[39] Loto RT and Loto CA (2012) Effect of P-phenylediamine on the corrosion of austenitic stainless steel type 304 in hydrochloric acid *Int. J. Elect. Sci.* **7(1)** 9423-9440.
[40] Loto RT, Leramo R and Oyebade B (2018) Synergistic combination effect of salvia officinalis and lavandula officinalis on the corrosion inhibition of low-carbon steel in the presence of SO_4^{2-} and Cl^- containing aqueous environment *J. Fail. Anal. Prev.* **18(6)** 1429-1438.
[41] Loto RT and Oghenerukewe E (2016) Inhibition studies of rosmarinus officinalis on the pitting corrosion resistance 439LL ferritic stainless steel in dilute sulphuric acid *Orient. J.Chem.* **32(5)** 2813-2832.

[42] Loto CA, Loto RT and Popoola API (2011) Effect of neem leaf (azadirachita indica) extract on the corrosion inhibition of mild steel in dilute acids *Int. J. Phy. Sci.* **6(9)** 2249-2257.
[43] Bockris JO and Swinkels DAJ (1964) Adsorption of n-decylamine on solid metal electrodes *J. of Elect. Soc.* **111(6)** 736.
[44] Vijayaraghavan K, Padmesh TVN, Palanivelu K and Velan M (2006) Biosorption of nickel (II) ions onto Sargassum wightii: Application of two-parameter and three parameter isotherm models *J. Hazard. Mater.* **B133** 304–308.
[45] Loto RT, Loto CA, Joseph O and Olanrewaju G (2016) Adsorption and corrosion inhibition properties of thiocarbanilide on the electrochemical behaviour of high carbon steel in dilute acid solutions *Results in Phys.* **6** 305-314. https://doi.org/10.1016/j.rinp.2016.05.013 (2016).

Finite Element Study on Forming Metal Sheets with an English Wheel

K J Fann

Department of Mechanical Engineering, National Chung Hsing University, 250, Kuokuang Road, Taichung 40227, Taiwan, R.O.C

E-mail: kjfann@nchu.edu.tw

Abstract. Because of the application trends of Industry 4.0, small-volume large-variety production becomes the daily life in the manufacturing industry. Techniques to flexibly shape sheet metal parts are therefore more and more inquired. This paper is thus aimed to have a preliminary glimpse at the shaping process made by the English wheel, which is broadly used by craftsmen, via Finite Element Analysis. As a result, the parts shaped by an English wheel with a zigzag path show a curved surface having a homogeneous curvature. The more the impression from the bottom roll onto the sheet metal, the smaller the radius of the bottom roll, and the thinner as well as the more compliant the sheet metal, the larger the curvature of the shaped metal part.

1. Introduction

Recently, more and more large sheet metal membrane covering made in curved surfaces are applied to modern architecture, either on the roofing or on the façade [1]. It thus needs a corresponding process to fabricate the sheet metal to precisely demonstrate the architect's design as well as construction demand. There are two types of traditional equipment, which can allow craftsmen to create such unusual or unique shape into metal sheets by flexible cold forming, the power hammer [2] and the English wheel [3]. The metal sheet is shaped by locally pushing the material aside, in that a compressive force is applied on the sheet metal to reduce the thickness directly either by impacting of the hammer or by rolling through the rolls. After the technique of the Incremental Sheet Forming (ISF) [4] is brought into the industry, a process combining ISF and impacting [5] along with an industrial robot has been introduced to more flexible shape the metal sheet as well. This encourages industry to extend the application of English wheel directly by using artificial neural networks in conjunction with an industrial robot [6]. However, the technique of the application of the English wheel is still not detailed. Thus, there is literature discussing the response on the sheet metal to the small impulse path of rolls. It is still worth to investigate the process parameters of the English wheel on the shape of the applied sheet metal, though. On this point of view, this paper is aimed to preliminarily investigate the parameters such as roll radius, impression, and path for a curved surface.

2. Setup of experiment and simulation

The process to flexibly cold shaping sheet metal parts with an English wheel can be illustrated in figure 1. A metal sheet is clamped with two rolls by lifting the bottom roll to impress the metal sheet. Via pushing or pulling the sheet metal to or away from the rolls, the rolls rotate because of the relative movement and the friction between the sheet metal and the rolls. A new impression onto the metal sheet is therefore continuously applied, where the metal is shifted between the rolls. The metal sheet will be dented and the material will spread out and thus become thinner there. Under different stress state along the thickness, the sheet will be bent and a curvature is thus created there.

To demonstrate the process of the flexible shaping metal sheet parts with the English wheel, this study created a finite element model according to the equipment made by Baileigh Industrial, which provides a throat depth of 750 mm, a flat upper roll having the width of 75 mm and a diameter of 250 mm, and several bottom rolls having the same width as the upper roll and the same height of 75 mm. In this study, three bottom rolls have been taken into account. They have the radii of 50 mm, 100 mm, and 200 mm, respectively. The metal sheet is a square having 500 mm each side. Two different thicknesses, 1 mm and 2 mm, were operated in the study. Two different metals, DC04 and A6061-T6, were used as well. Two impressions, a half and a quarter of the thickness, were developed by lifting the bottom die into the sheet metal. Three different paths were applied in this study, as shown in figure 2. An angle of 4.5° for case (a) the sawtooth zigzag path or 9° for case (b) the triangle zigzag was applied on the paths, so that the distances between turn points are the same for case (c) the rectangle zigzag path. There are some spaces between the contact center and the edge of the sheet metal. In this study 20 mm is set.

Figure 3 shows the typical model used in this study. The metal sheet is modelled with shell elements, which are in square shape having the mesh size of 5 mm. The top and bottom rolls are modelled with square shell elements as well. Each element has a central angle less than 4.5°, so that the geometry detail of the rolls can be revealed accordingly. However, they are regarded as rigid. 0.15 has been set for the friction coefficient between the sheet metal and the rolls. The Finite Element Analysis was conducted with the commercial explicit code LS-DYNA.

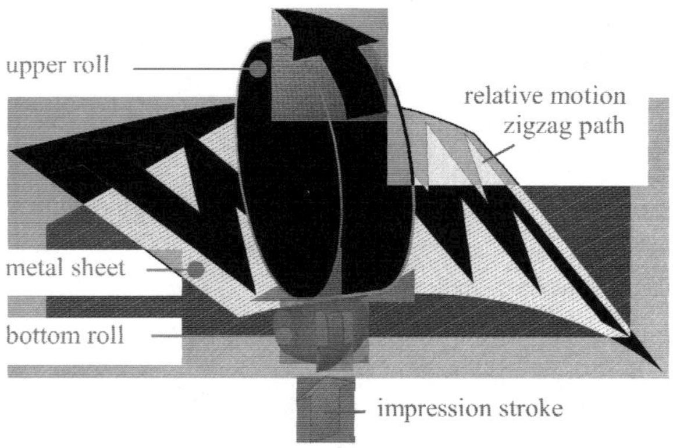

Figure 1. Illustration of the flexible shaping process of a sheet metal with the English wheel.

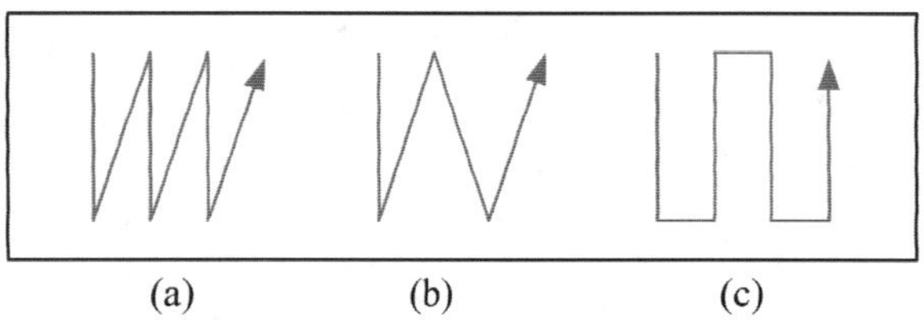

Figure 2. Zigzag process paths used in this study: (a) sawtooth, (b) triangle, and (c) rectangle.

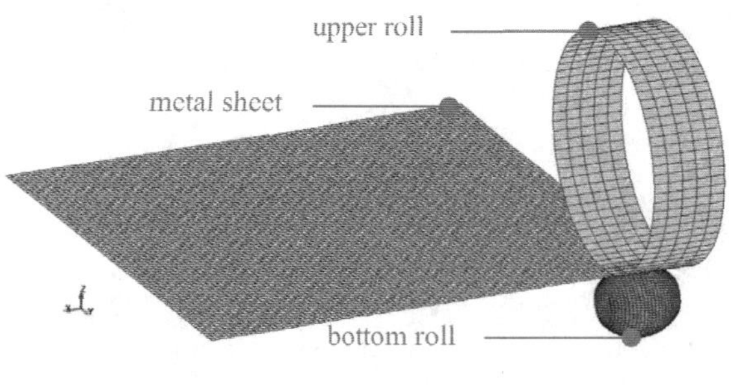

Figure 3. Finite Element Analysis Model.

3. Results and Discussion

Figure 4 shows that the effective plastic strain obtained on the final shape of a DC04 sheet in thickness of 1 mm out of the sawtooth zigzag path impressed by the English wheel having the bottom roll in 25 mm of radius with an impression of 0.5 mm. It can be observed that the part is crowned and the area having a significant plastic strain conforms to the process path. However, not on all the surface covered by the rolls a significant plastic strain has been achieved. It seems that the process has a dominant effect on the areas around the turning points, while a less effect is acted on the middle of the path between the turning points. has. The inhomogeneous plastic strain distribution might cause the fluctuation shown in figure 5, which shows the curvature radius profile on the sliced cross-sections defined in figure 6 and calculated according to Modified Least Square Method [7] for every 25 mm or every 5 elements along the cross-sections. It can be seen, that the radius along the cross-sections at the path middle point (Y = 239 mm) fluctuates not so much as those calculated at the turning points (Y = 4 mm and 438 mm). The average radius at the path middle point is 505 mm, while 921 mm and 727 mm are read for the turning points at 40 mm and 438 mm, respectively.

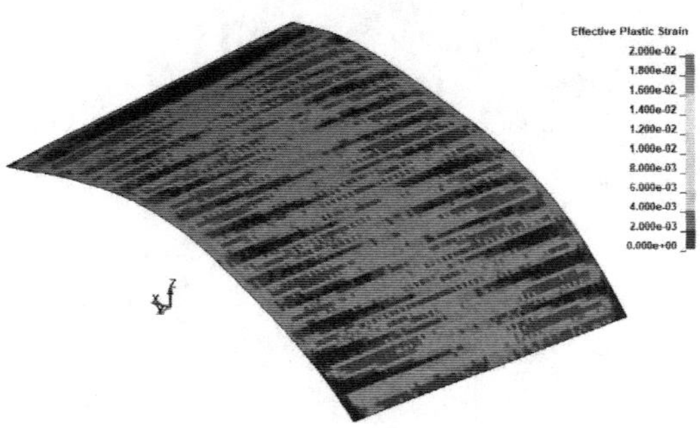

Figure 4. Effective plastic strain on the shaped sheet metal parts shaped with the sawtooth zigzag path.

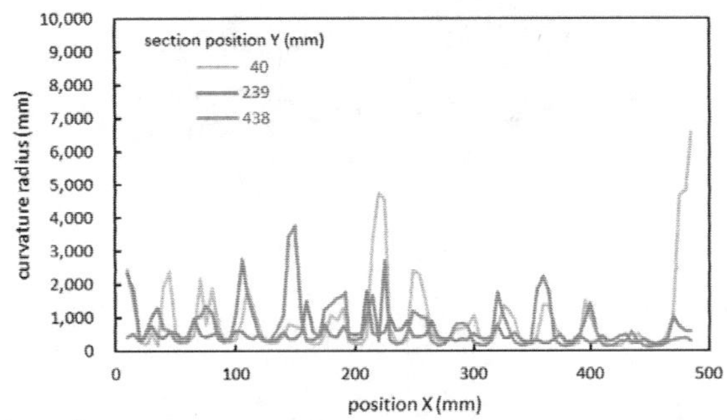

Figure 5. Calculated curvature radius along the cross-sections sliced perpendicular to the process path.

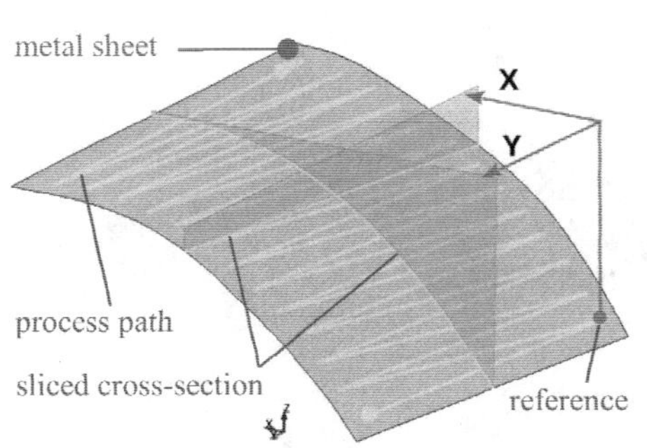

Figure 6. Position of the sliced cross-sections defined for the curvature radius calculation.

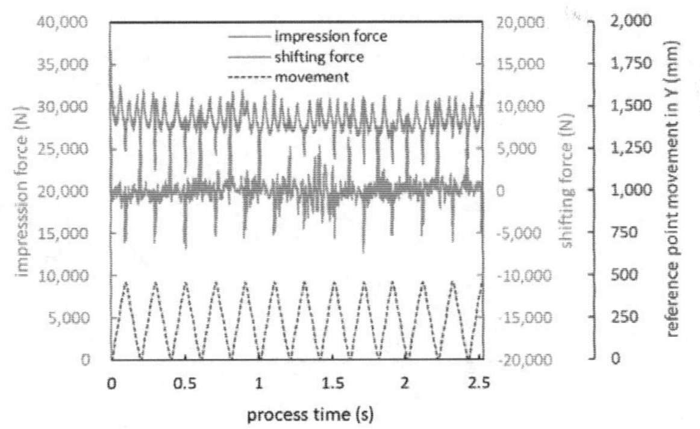

Figure 7 Reaction force on the sheet metal in the thickness direction (impression force) and on the panel (shifting force) under the sawtooth zigzag path.

The reaction force on the metal sheet can be regarded as the force needed for the process as shown in figure 7, in which the relative movement of a reference point on the metal sheet to the rolls shown at the bottom area of figure 7 is used as reference to indicate the state of the process. There are two kinds of reaction forces, the impression force and the shifting force. The impression force comes in the thickness direction of the metal sheet and is reacted to the both rolls because of the impression of the bottom roll into the sheet metal. The shifting force is formed on the sheet panel by pushing (positive shifting force) or pulling (negative shifting force) the metal sheet to build up the process path. Peaks of the reaction forces can be observed at the turning points of the process path. It might be caused by the acceleration and the deceleration during shifting and rotating the metal sheet around the turning points. Currently, the shifting velocity is set at 5,000 mm/s and the rotating speed is set at 131 rad/s to reduce the process (computing) time.

Similar results can be found in the other processes under the above mentioned parameter settings. For example, figure 8 shows that the effective plastic strain on the sheet parts shaped with the triangle zigzag path, while that with the rectangle zigzag path is shown in figure 9. The area covered with the significant plastic strains conforms to the process path again. Both of the shaped sheet metal parts are flatter than that shaped by the sawtooth zigzag path as shown in figure 4. It can be attributed that the area covered by the significant plastic strains are smaller and sparser.

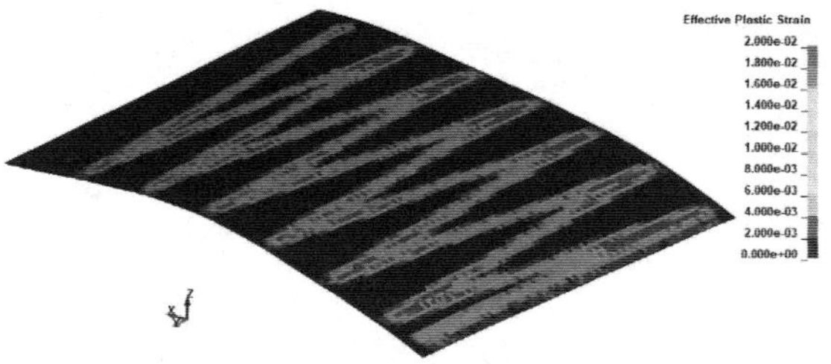

Figure 8. Effective plastic strain on the sheet metal part shaped with the triangle zigzag path.

Figure 9. Effective plastic strain on the sheet metal part shaped with the rectangle zigzag path.

Figure 10 shows the curvature radius calculated from the sliced cross-sections perpendicular to the process path with the above mentioned parameter settings. Some results from three zigzag process paths have been shown in Figure 4 to 9. However, regarding the curvature radius, the most densely covered sawtooth path has the smallest radius than the sparsely covered triangle and rectangle paths.

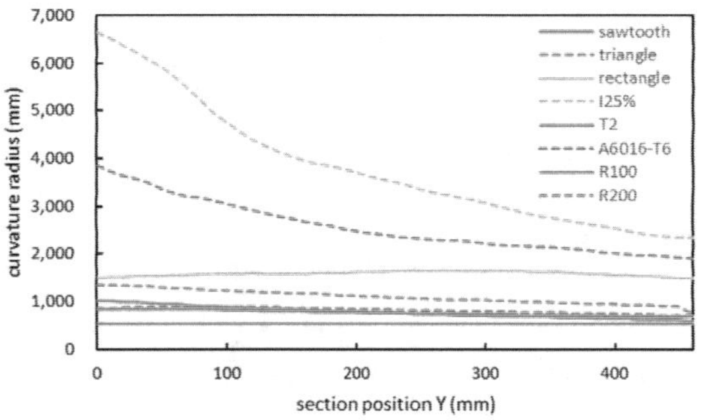

Figure 10. Calculated curvature radius on the sliced cross-sections of the sheet metal parts shaped with different parameter settings.

Figure 11. Effective plastic strain on the sheet metal part shaped by an impression in one quarter of the thickness.

A further process setting without replacing the tools or the sheet metal is to reduce the impression of the bottom roll to a quarter of the sheet thickness, which is shown in dashed blue line with the legend "I25%" in Figure 10 and has the largest curvature radius of all shaped sheet metal parts. Figure 11 shows this shaped sheet metal part fringed with the plastic strains in a scale just one tenth used in the other Figures. That means that only few plastic strain is carried out on the sheet metal part and thus a marginal curvature is brought on the part.

However, if a thicker metal sheet, such as 2 mm (with the legend "T2" in Figure 10), is shaped by the English wheel with the same parameter settings as before under an impression of a half of the thickness, the curvature radius brought into the shaped sheet metal is not smaller than that of 1 mm, even severe plastic strains are already covered onto the sheet part. That means that the plastic strain brought into the sheet metal might be a dominant but not the only one parameter on the curvature brought into the sheet metal part shaped by the English wheel. A thicker sheet metal might have some stiffer characteristics than a thinner sheet, so that a further curvature might not be easily brought into the thicker metal sheet.

If an aluminum alloy sheet A6061-T6 is shaped with an English wheel, only a remarkably small curvature radius onto the part can be achieved in comparison to the steel sheet DC04, as shown in Figure 10. Less plastic strain is carried out than on the steel sheet shaped with an impression of one quarter of the thickness as shown in Figure 11. Only few plastic strain is brought onto the aluminum alloy sheet can be attributed to that the aluminum alloy has a Young's modulus only one third of steel. For a relative totally small amount of the strain impressed by the rolls, the elastic part of the strain is not negligible in comparison to its plastic part. Thus, for a certain total strain, the lower the Young's modulus, the more the elastic part of the strain. The wrinkling shown on the side along the last process path might be caused by the numerical error from hourglass energy, which is no significant influence on the result regarding to the curvature of the shaped sheet part.

If the impression roll is switched with a roll having another radius like 100 mm or 200 mm instead of 50 mm, a larger curvature radius can be obtained on the steel sheet part, as shown in Figure 10 with the legends "R100" and "R200." There is no significant difference on the curvature radius brought by these two rolls. The width of the stripes covered with the significant plastic strain along the process path is wider than that shaped with a roll having a diameter of 50 mm as shown in Figure 4. Nevertheless, the curvature radius of the shaped part is larger, even more area of the shaped sheet metal part covered with the significant plastic strain. This is the other evidence for the above mentioned statement, that the plastic strain achieved on the sheet metal might not be the only one parameter on the curvature shaped by the English wheel. It might be concluded that the sharper the indentation or the impression is, the smaller the curvature radius can be achieved.

4. Conclusion

This paper has preliminarily demonstrated the flexible cold sheet metal shaping process carried out by the English wheel, which is broadly used by craftsmen for making creative surfaces of the metal sheet, via Finite Element Analysis with an explicit code LS-DYNA. Study on the curvature of the shape metal part has been conducted in this paper by using different process parameter settings, such as the process path, the thickness, the material, the impression, and the geometry of the impression roll.

As a result, more curvature can be brought into the sheet metal part by using the sawtooth zigzag path than by using the triangle or rectangle zigzag path. The more impression applied to the metal sheet from the bottom roll, the more curvature can be carried out in the part. The smaller the diameter of the impression roll, the more curvature can be achieved as well. However, the thinner and the stiffer the sheet metal, the more curvature can be realized onto the shaped sheet metal part.

References

[1] Ching F D K 2015 *Architecture – Form, Space, & Order* (Hoboken: John Wiley & Sons)
[2] Fournier R and Fournier S 1990 *Metal Fabricator's Handbook* (Los Angeles: Price Stern Sloan)
[3] Barr E 2019 *Sheet Metal Shaping: Tools, Skills, and Projects* (Beverly: Motorbooks)
[4] Jeswiet J, Micari F, Hirt G, Bramley A, Duflou J and Allwood J 2005 *CIRP Ann. – Manuf. Technol.* **54** 2 pp 88–114
[5] Tanaka H, Naka S and Asakawa N 2012 *Int. J. Automot. Technol.* **6** 5 pp 604–610
[6] Rossi G. and Nicholas P 2018 *Proc. 38th Ann. Conf. of the Association for Computer Aided Design in Architecture*
[7] Umbach U and Jones K N 2003 *IEEE Trans. Instrum. Meas.* **52** 6 pp 1881–1885

IOP Conf. Series: Materials Science and Engineering 1222 (2022) 012007 doi:10.1088/1757-899X/1222/1/012007

Vibration Fatigue Analysis of Interface Enclosure with Cantilever Structure for Charging Commercial Electric Vehicles

Chan Hee Park[1,a], Shin Ho Ryu[1,b], Jung Dam Cha[2,c] and Gwang Ho Kim[2,d] *

[1]PROTECHKOREA Co. Ltd., 120-3, Nanosandan-ro, Samtae-ri, Nam-myeon, Jangseong-gun, Jeollanam-do, Republic of KOREA

[2]Dept. of Mechanical System, Gwangju Campus of Korea Polytechnics, 85, Haseo-ro, Buk-gu, Gwangju, Republic of KOREA

[a]chanhui6332@daum.net, [b]chosun00@hanmail.net, [c]cjd7909@kopo.ac.kr, [d]ghkim@kopo.ac.kr (corresponding author)

Abstract. The interface enclosure for charging commercial electric vehicle (CEV), which is an alternative to the fuel filler of internal combustion engine vehicle, was studied. It supplies the power and the communication signal for charging between the power supply (EVSE) and the battery. It should be long to match the exterior of the vehicle for easy charging, and it requires space between the vehicle frame and the enclosure to connect the charging cable. In order to design the interface enclosure with the cantilever structure which is vulnerable to vibration, analysis-based design through vibration fatigue analysis was conducted, and the reliability of analysis results was reviewed through experiments.

1. Introduction

Recently, the need for zero emission vehicles (ZEVs) is increasing due to the strengthening of greenhouse gas (GHG) regulations and the seriousness of particulate matter (PM). In particular, carbon dioxide emissions of trucks, which account for most road freight transport, have more 2.5 times than ones of passenger cars. Therefore, it is essential to develop the commercial electric vehicle (CEV). As the battery price, which has burdened the vehicle price, continues to decline, commercial electric vehicles are being developed mainly for vans, buses, and logistics service vehicles that do not have long distances per day. The electric vehicle supple equipment (EVSE) monitors the charging environment through communication (PLC or CAN) with the battery electric vehicle (BEV) and provides power to charge the battery through charging inlet. When converting an internal combustion vehicle (ICV) to an electric vehicle (EV), passenger cars and commercial vehicles of 1 ton or fewer install the charging inlet to the vehicle frame. However, medium and heavy trucks require additional parts to mount the charging inlet to the understructure frame. As the demand for electric vehicles increases, the need for high-capacity and rapid charging is increasing. Structural stability of the enclosure to protect the charging circuit from the harsh road driving environment is becoming more important for high voltage charging. In this paper, the interface enclosure that is installed to the lower frame of a commercial electric vehicle for protecting the inlet circuit was studied. It is designed to stably protect the circuit inside the enclosure by various vibrations generated while driving the vehicle. Vibration fatigue analysis, which performs durability analysis in the frequency domain, was used to secure sufficient mechanical stability of the interface enclosure with the cantilever structure, which is vulnerable to vibration. The reliability of the analysis results was verified through the resonance-point search test, the experimental stress analysis, and the random vibration test.

2. Results and Discussion

2.1. *Vibration fatigue analysis*

Random vibration in mechanical engineering means that future behavior cannot be accurately predicted with non-deterministic motion. Vibrations generated by vehicles on the road, rocket launches, and airplane wings during turbulence are not repetitive and unpredictable [1,2]. The data acquired through the accelerometer is difficult to predict and not repeatable. Therefore, a random vibration test is needed to reflect the reality. Structural responses to random vibrations are generally handled using statistical or probabilistic approaches. Random vibration is more realistic than sinusoidal vibration because it excites not only the resonant frequency but also all frequencies at the same time. FFT is excellent for analyzing vibrations when there is a finite number of dominant frequency components, whereas power spectral density (PSD) is used to characterize arbitrary vibration signals [3]. The random vibration analysis uses the modal superposition technique as the basis for the harmonic response. It is scaled and combined to the input power spectral density to obtain the dynamic response and the equivalent PSD projected onto the critical plane [4]. The vibration exciter applies random accelerations over a bandwidth of approximately 20 to 2000 Hz [5]. Most signals have a bell-shaped, symmetrical probability density function(PDF). Gaussian signal has 3 kurtosis with a probability of 99.73% at 3σ. The vibration fatigue analysis is estimated through the process shown in figure 1.

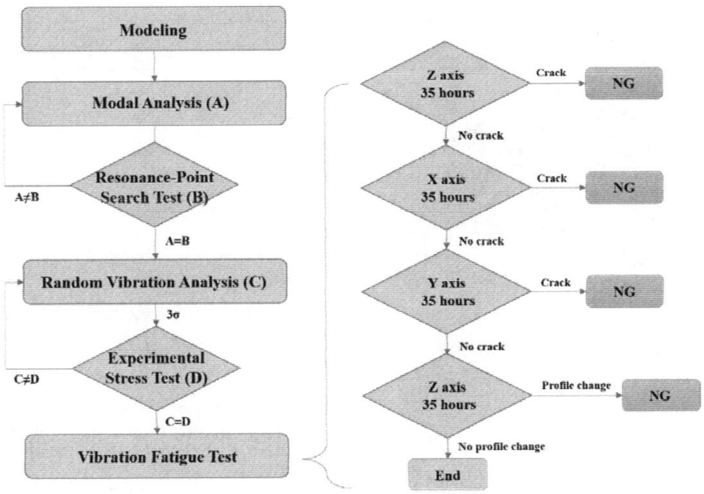

Figure 1. Vibration fatigue analysis process

2.2. *Modal analysis*

Natural frequencies and mode shapes are important parameters that characterize the dynamic responses of test parts during vibration test [6]. The model of interface enclosure for modal analysis as shown in figure 2 was simplified. It was set to match weight of the actual enclosure using a distributed mass. In addition, simulation conditions for bolt connection were implemented by using point mass. It was analyzed up to 10th mode using the pre-stress condition. To verify the result of the modal analysis, it

IOP Conf. Series: Materials Science and Engineering 1222 (2022) 012007 doi:10.1088/1757-899X/1222/1/012007

was compared with the resonance-point search test of figure 3. Table 1 shows the simulation results and vibration test results. The 1st mode shape in the X-axis direction was predicted at 44 Hz, and the fundamental natural frequency measured by the resonance-point search test was 42 Hz. Analysis results were similar to vibration test results.

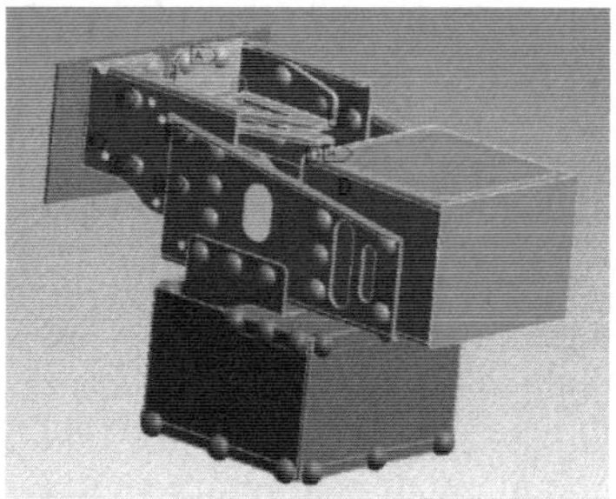

Figure 2. Modeling for modal analysis

Figure 3. Setup for Resonance-point search test

Table 1. compare modal analysis with resonance-point search test

	Modal analysis	**Resonance-point search test**
x-axis (Fr-Rr)	 1st mode 44.0Hz	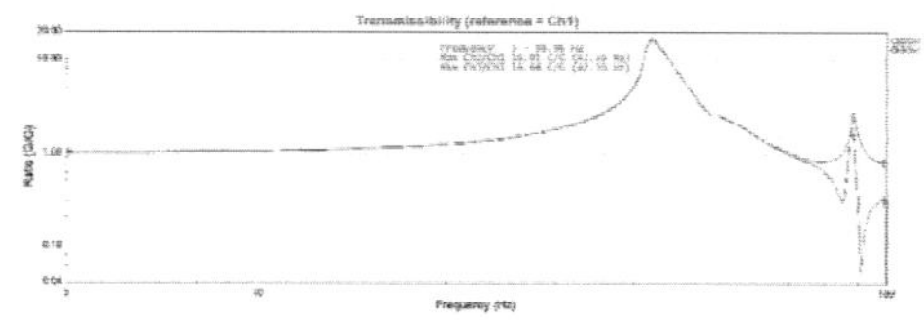 42.56Hz, 16.9G/G
y-axis (Rh-Lh)	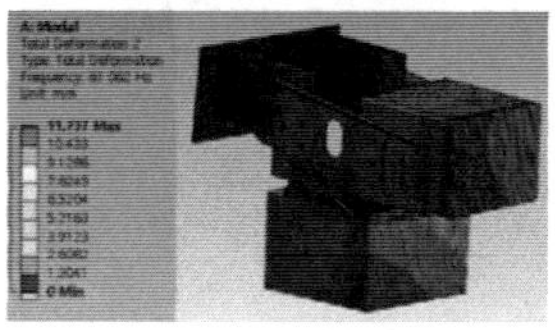 2nd mode 61.1Hz	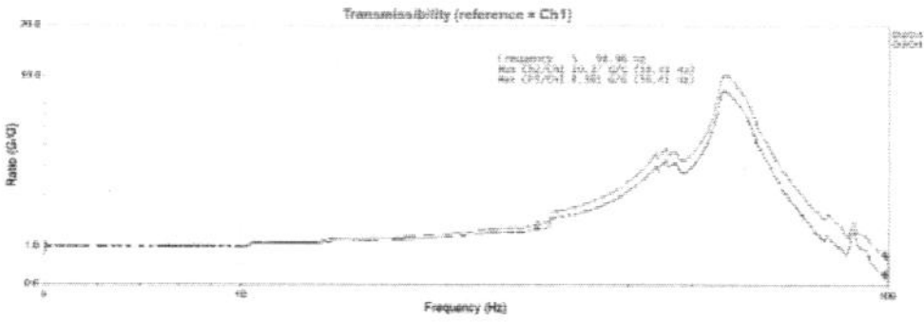 56.4Hz, 10.3G/G
z-axis (vertical)	not present within 100Hz	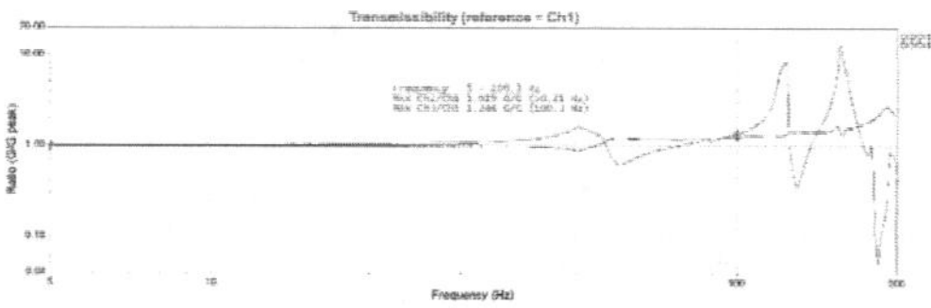 100.3Hz, 9.9G/G

2.3. Random vibration analysis

The purpose of random vibration analysis is to determine the response of a structure to a random vibration load. The frequency of the time history is captured along with statistics and used as a load in random vibration analysis. For historical reasons, this spectrum is referred to as the power spectral density. The load profile of input in the x-axis (front-rear direction of the vehicle), y-axis (left-right direction of the vehicle), and z-axis (up-down direction of the vehicle) was power spectral density. A

IOP Conf. Series: Materials Science and Engineering 1222 (2022) 012007 doi:10.1088/1757-899X/1222/1/012007

solution of modal analysis is coupled with the initial conditions for random vibration analysis. The simulation results were evaluated at the 3 sigma level, and were compared with results acquired by attaching a Rosette strain gauge to the surface of the charging interface enclosure. As shown in figure 4, the signals of strain gauge were combined for the three axes, and the stress were converted. The results of experimental stress test were compared with the results of random vibration analysis. Test results were similar to simulation results. Based on the results of random vibration analysis, a vibration fatigue test was conducted. Each axis was excited for 35 hours and the cracks were checked. If there was no crack in all axes, an additional vibration test was conducted on the z-axis for 35 hours. Changes in the vibration profile were examined, such as increase in amplitude or shift of resonance frequency, etc. It was confirmed that the vibration profile was not changed during test as shown in figure 5, and bolts were not loosed.

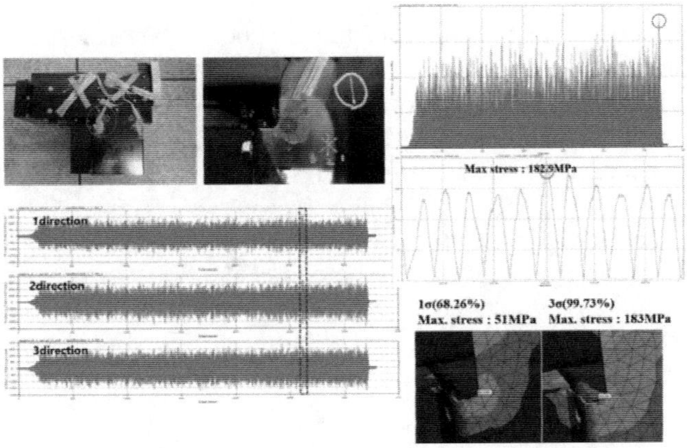

Figure 4. Compare strain gauage test with random vibration analysis

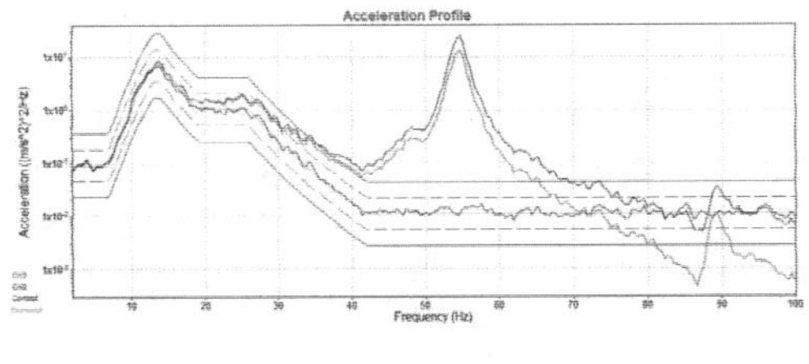

(a) Vibration profile after 2hr

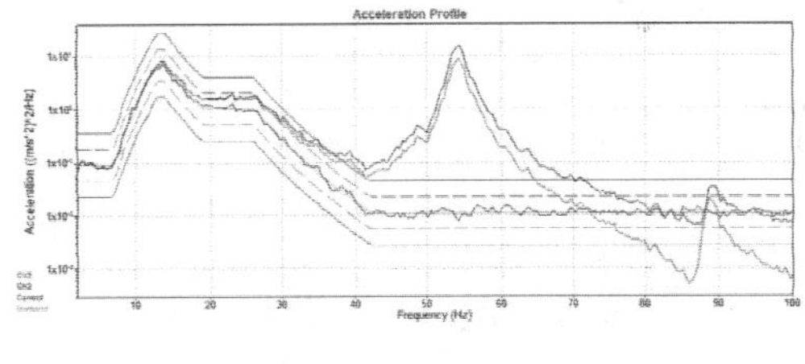

(b) Vibration profile after 10hr

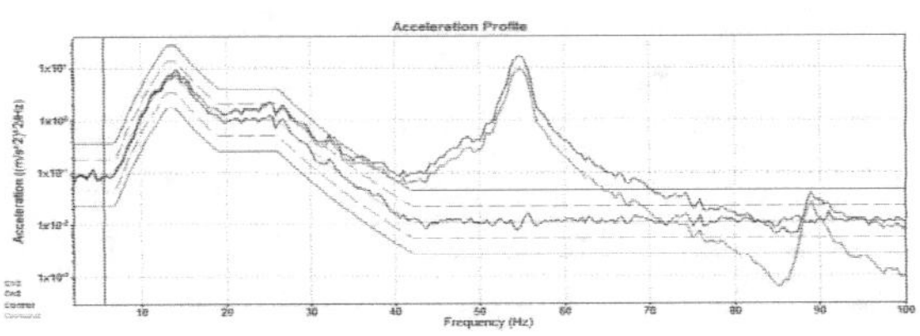

(c) Vibration profile after 20hr

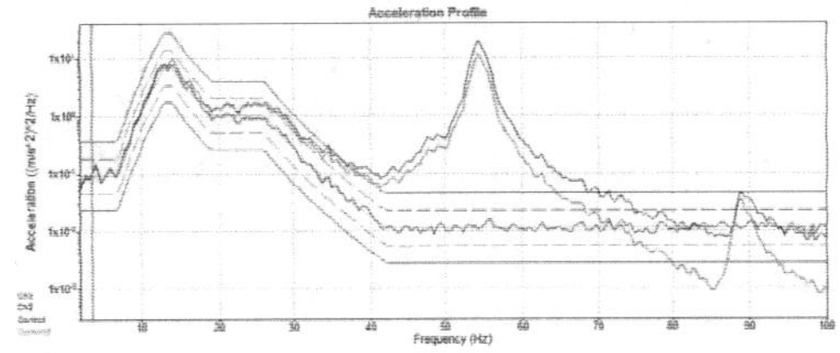

(d) Vibration profile after 35hr

Figure 5. Vibration profile according to additional Z-axis excitation in vibration fatigue test

3. Conclusions

Interface enclosure was designed for charging commercial electric vehicles. It has a cantilever structure due to design constraints. It was designed to have sufficient vibration durability through vibration fatigue analysis, and the reliability of analysis results was reviewed through the resonance-point search test, the experimental stress analysis, and the random vibration test.

1. Considering the PSD data of random vibration, the fundamental natural frequency of enclosure should be designed higher than 40Hz to avoid severe resonance. Through modal analysis, the enclosure was designed so that the first mode shape occurred above 40Hz. As a result of the analysis, the first mode shape occurred at 44 Hz. The fundamental natural frequency of 42 Hz was confirmed in the resonance-point search test.

2. Random vibration analysis was performed for the fatigue analysis of structures caused by external excitation. Since the time history deviating from $\pm 3\sigma$ is probabilistically within 0.3%, the equivalent stress was estimated using the $\pm 3\sigma$ scale factor. In order to evaluate the results of the vibration fatigue analysis, an experimental stress analysis was performed using a strain gauge. It was confirmed that the analysis result reflects the measured stress result well.

3. A charging interface enclosure with cantilever structure was designed through analysis-based design. As a result of the random vibration test, there was no damage to the interface enclosure and no loosening of bolts was found. Reliability was verified by comparing analytical and experimental techniques. It is expected that reliable design can be secured by applying the vibration fatigue analysis when design changed in the future.

Acknowledgments
This research was financially supported by the Ministry of Small and Medium-sized Enterprises(SMEs) and Startups(MSS), Korea, under the "Regional Specialized Industry Development Plus Program(R&D, S2837705)" supervised by the Korea Institute for Advancement of Technology(KIAT).

References

[1] Hu J 1995 *J. Ind. Electron. Soc.* **38** 34-40
[2] Li R S 1999 *J. Electron. Packag.* **123** 394-400
[3] Lin J S and Yim K S 2000 Application of random vibration test methods for automotive subsystems using power spectral density (PSD) *SAE Tech. Pap. 2000-01-1331*
[4] Teixeira G M, Jones D and Draper J 2014 Random vibration fatigue-A study comparing time domain and frequency domain approaches for automotive applications *SAE Tech. Pap. 2014-01-0923*
[5] Lambert R G 1976 *Shock Vib. Bull.* **46** 55-72
[6] Yu D, Al-Yafawi A, Park S and Chung S 2010 *Proc. 60th Electron. Comp. Techn. Conf. (ECTC): IEEE* 188-193

Effects of directional solidification on hardness and microstructure in Al–3wt.%Cu–1wt.%Li alloy

G A dos Santos[1,2,a], T L Alves[1,b], R Teram[1, 2,c], R A Cruz[1, 2,d], M S Nascimento[1,3,e], V T dos Santos[2,4,f], M R da Silva[4,5,g], A A Couto[2, 3,h] and I F Machado[5,i]

[1] Federal Institute of Education, Science and Technology of São Paulo, São Paulo, SP, Brazil
[2] Mackenzie Presbyterian University, São Paulo, SP, Brazil
[3] Institute of Nuclear and Energy Research, IPEN, São Paulo, SP, Brazil
[4]Termomecanica São Paulo S.A., São Bernardo do Campo, SP, Brazil
[5]Polytechnic School, University of São Paulo, SP, Brazil

[a]givanildo@ifsp.edu.br, [b]alves.tiago@aluno.ifsp.edu.br, [c]rogerioteram@ifsp.edu.br, [d]ricardoapcruz@ifsp.edu.br, [e]mauricio.nascimento@ifsp.edu.br, [f]vinicius.santos@termomecanica.com.br, [g]marcio.rodrigues@termomecanica.com.br, [h]antonioaugusto.couto@mackenzie.br, [i]machadoi@usp.br

Abstract. Solidification is probably one of most important phase transformations in materials manufacturing. Additionally, a wide range of solidification microstructural features is obtained depending on processing parameters. Mechanical properties of the cast are influenced by the microstructure obtained after solidification. The aim of this work is to investigate how solidification conditions affect solidification and the resulting microstructure of the Al–3wt.%Cu–1wt.%Li (ternary system) alloy. The alloy was solidified unidirectionally upward through a water-cooled steel plate. Results include secondary dendritic arm spacing (SDAS), hardness and microstructure analysis using optical microscopy. Results showed that SDAS values tended to increase as the distance from the heat-exchange surface increased (P). Conversely, hardness decreases as P increased.

1. Introduction

Properties such as low density and higher modulus of elasticity than conventional alloys make aluminum-lithium alloys attractive for aerospace applications. Li additions enable the formation of effective hardening precipitates and provide higher fatigue-crack growth resistance [1,2]. The addition of lithium and copper is promising as it allows for an attractive combination of properties, including a high strength-to-weight ratio, good machinability, environmental friendliness, and elevated fatigue strength, which determines the prospects of their application in aircraft structures [3,4].
Casting is an example of a manufacturing solidification process, and many variables must be. Thus, several authors addressed their studies on the resulting microstructure influence on the properties of engineering materials [5,6,7,8]. Generally, structures with more refined grains tend to increase the strength of the metallic material, as described by the well-known Hall-Petch equation [9,10]. However, for cast metals and alloys, this relationship is not always observed, as the effect of grain size can be overcome by increasing the number of imperfections (micro porosities), increasing the percentage volume fraction of the second phase or arm spacing dendritic [11].

The aim of the present work was to investigate how solidification conditions affect the microstructure of the Al–3wt.%Cu–1wt.%Li (ternary system) alloy. Solidification was carried out by heat extraction. This alloy was solidified unidirectional upward. Results include secondary dendritic arm spacing (SDAS), hardness, and microstructure analysis by optical microscopy.

2. Materials and Methods

Figure 1 shows the experimental apparatus. It can be observed that heat was extracted directionally through a low carbon steel plate (SAE 1020) with a thickness of 3 mm. A split ingot mold with an outer diameter of 70 mm, an inner diameter of 60 mm and a height of 157 mm was used. The internal surfaces of the ingot mold were coated with layers of alumina.

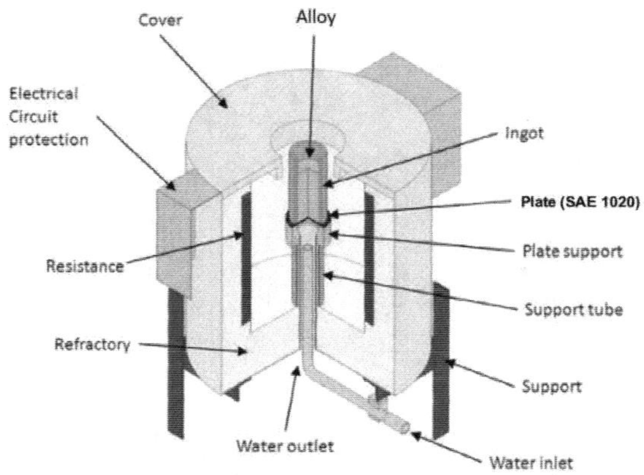

Figure 1. Unidirectional solidification furnace [12].

The aluminium alloy was prepared in a graphite–clay crucible placed in a muffle furnace using a commercial AA 2198 alloy. The alloy was heated to a temperature of 10°C above the liquidus temperature and then poured into the split ingot mold inside the solidification furnace [1]. Prior to casting, the solidification furnace was heated to reach the alloy-casting temperature. The heat-exchange surface was polished to decrease heat-transfer resistance. Type K thermocouples of 1.6 mm in diameter were placed at distances of 4, 8, 12, 22, 52, 68 and 88 mm from the plate from which the heat is extracted. All thermocouples were connected to a data logger interfaced with a computer, with a data acquisition interval equal to 1 s. The cooling system was activated after all thermocouples indicated temperatures of approximately 660 °C.

Selected cross sections (perpendicular to the heat extraction direction) were taken from the cast ingot at different distances from the heat-extraction plate, polished and etched. For metallography preparation, Tucker´s reagent was used (a solution of 45 mL HCl, 15 mL de HNO_3, 15 mL HF, and 25 mL distilled water). Image-processing system Zeiss AxioVert A1 microscope (Carl Zeiss, Gottingen, Germany) was used to measure SDAS (about 30 independent readings for each selected position) and their distribution range. The Zeiss AxioVert microscope (Carl Zeiss, Göttingen, Germany) was used to measure SDAS (about 30 independent readings for each transversal sample). Hardness values were obtained on a Wilson UH-930 hardness tester (Boehler, Lake Bluff, IL, USA) in accordance with ASTM E10-2012 [13]. A load of 62.5 kgf and a sphere 2.5 mm in diameter were used. Five hardness tests were performed on each sample. Graphs and experimental equations were obtained with the SciDAVis software version 1.25.

3. Results and Discussion

Figures 2 and 3 presents the mean experiment values of SDAS as a function of distance to P and as function of tip-growth rate (V_L). SDAS values tended to decrease to positions close to the heat-extraction

IOP Conf. Series: Materials Science and Engineering 1222 (2022) 012008 doi:10.1088/1757-899X/1222/1/012008

plate. These positions had higher V_L values, suggesting that higher cooling rates imply a more refined microstructure.

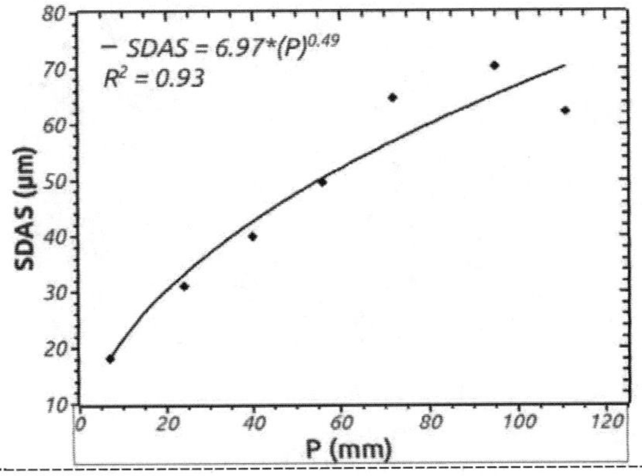

Figure 2. Correlation between secondary dendrite arm spacing (SDAS) and distance from heat-extraction plate (P).

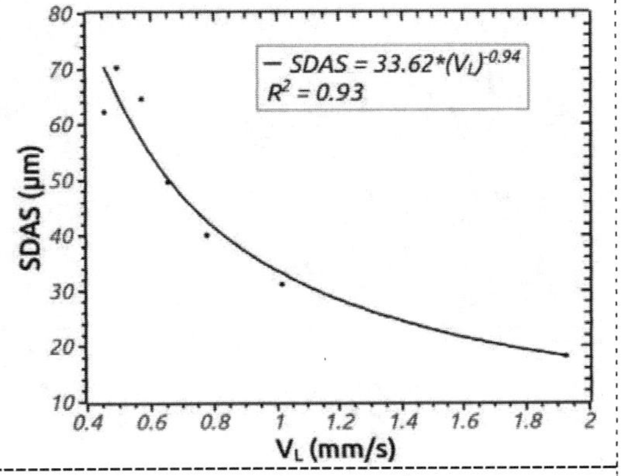

Figure 3. Correlation between secondary dendrite arm spacing (SDAS) and tip-growth rate (V_L).

Obtained microstructures along the longitudinal section of the Al-3wt.%Cu-1wt.%Li alloy are shown in figures 4, 5, 6 and 7. The microstructures were obtained by optical microscopy at positions 7, 24, 56 and 111 mm from the heat-extraction plate. It was observed that the size of the grains increases towards positions more distant from the heat exchange plate.

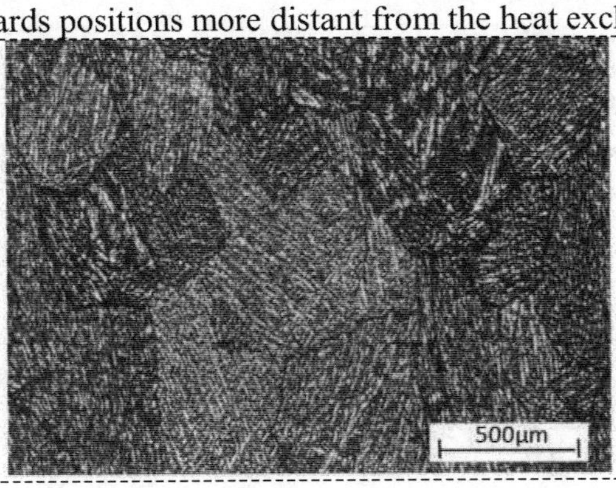

Figure 4. Micrograph of the sample taken from the 7 mm position as a function of the heat-extraction plate.

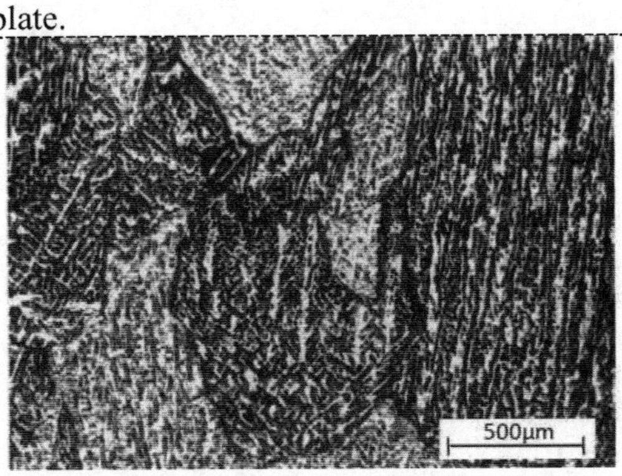

Figure 5. Micrograph of the sample taken from the 24 mm position as a function of the heat-extraction plate.

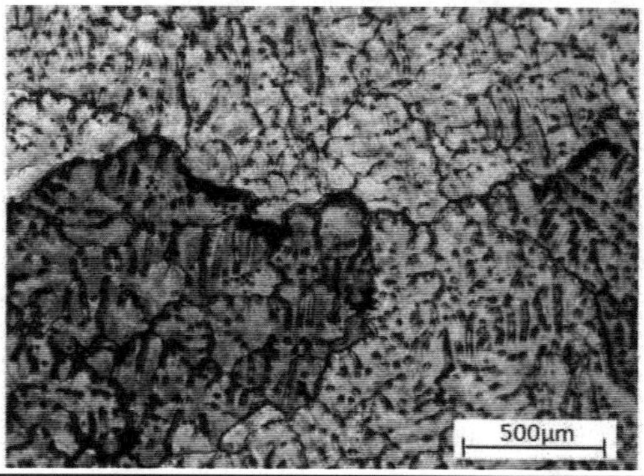

Figure 6. Micrograph of the sample taken from the 56 mm position as a function of the heat-extraction plate.

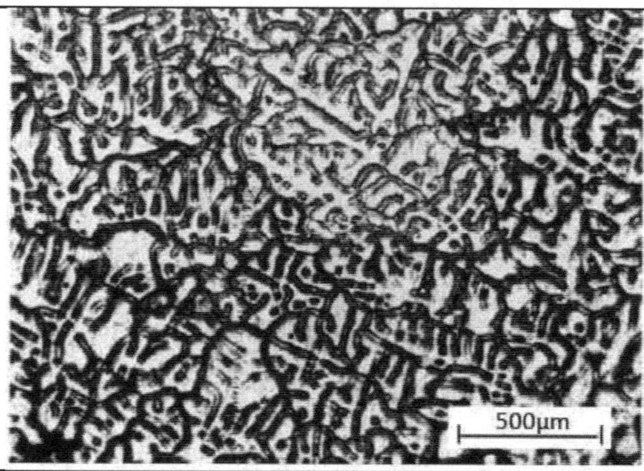

Figure 7. Micrograph of the sample taken from the 111 mm position as a function of the heat-extraction plate.

The data obtained in the hardness test are shown in figures 8 and 9. The hardness values (HB) were correlated with the values of distance from the heat exchange plate (P) and with the SDAS values. The experimental equations were obtained by the method of least squares. Error bars were defined by the standard deviation of the experimental data. The linear fit of the data suggests that HB values decrease with increasing distance from P and SDAS values.

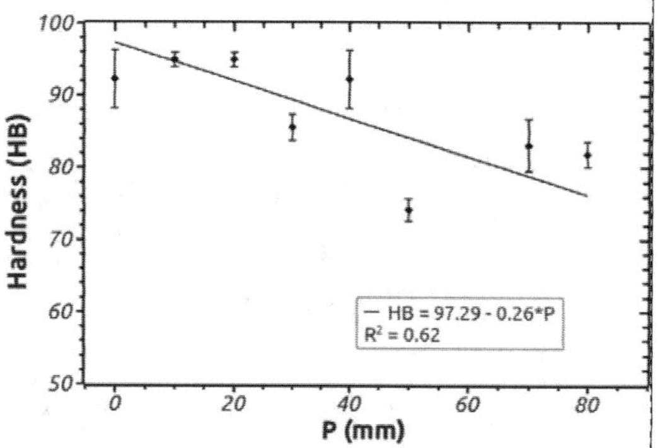

Figure 8. Correlation between hardness (HB) and distance P.

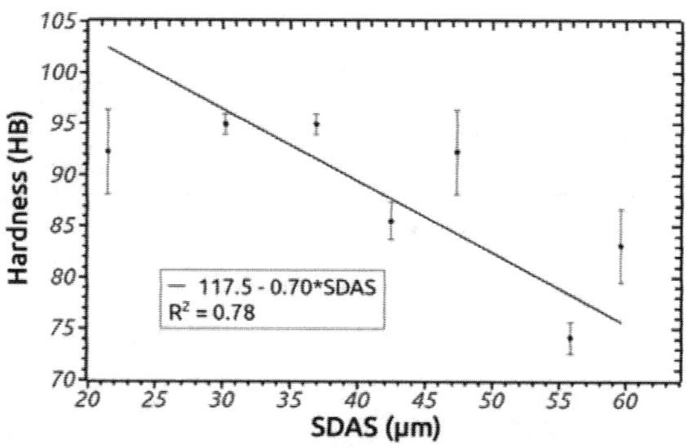

Figure 9. Correlation between hardness (HB) and SDAS.

4. Conclusions

We can draw the following conclusions:

(1) SDAS values decreased as distance from P increased.
(2) The structure is more refined in regions closer to the heat exchange surface.
(3) The experimental equations obtained indicate that HB decreases with increasing distance from the heat-extraction plate.
(4) The experimental equations obtained indicate that HB values decrease with increasing SDAS.

Acknowledgments

Authors acknowledge the support received for research activities: CAPES, Educational Center of the Salvador Arena Foundation, Termomecanica São Paulo S.A., Mackenzie Presbyterian University, Energy and Nuclear Research Institute (IPEN) and Federal Institute of Education, Science and Technology of São Paulo (IFSP).

References

[1] Santos GA, Goulart P.R., Couto A.A., *et al.*, "Primary dendrite arm spacing effects upon mechanical properties of an Al-3w%Cu-1w%Li alloy", In: Öchsner, A., Altenbach, H. (eds), *Properties and characterization of modern materials*, Springer, 2017. Available at: https://doi.org/10.1007/978-981-10-1602-8_19

[2] Rioja R J, Liu J. The evolution of Al-Li base products for aerospace and space applications. *Metallurgical and Materials Transactions A*. **43** 3325, 2012.

[3] Du J, Zhang A, Zhang L, Xiong S, Liu F. Quantitative and qualitative correlations by atomistic determination for the precipitated phases in Al–Li-Cu system. *Intermetallics* **112** 106551, 2019.

[4] Kraposhin V S, Kolobnev N I, Ryabova E N, Everstov A A, Talis A L. Inhomogeneous Solid Solutions in Alloys of the Al – Cu – Li System: Possible Structure of Clusters. *Metal Science and Heat Treatment* **61** 73, 2019.

[5] Nascimento MS, Frajuca C, Nakamoto FY, Santos GA, Couto, AA. Correlação entre variáveis térmicas de solidificação, microestrutura e resistência mecânica da liga Al-10%Si-2%Cu. *Matéria* **2017**, *22*, e11774. Available at: https://doi.org/10.1590/S1517-707620170001.0106

[6] Nascimento MS, Franco ATR, Frajuca C, Nakamoto FY, Santos GA, Couto AA: Materials Research Vol. 21 (5) (2018). Available at: https://doi.org/10.1590/1980-5373-MR-2017-0864

[7] Santos GA, Ribeiro AN, Nascimento MS, Frajuca C, Nakamoto FY, Silva MR, et al. Influence of the thermal parameters on the microstructure, corrosion resistance and hardness on the unidirectional solidification of al-10wt% si-5wt% cu alloy. Mater Sci Forum 2020;1012 MSF:308-313. Available at: https://doi.org/10.4028/www.scientific.net/MSF.1012.308

[8] Cruz R A, Santos G A, Nascimento M S, Frajuca C, Nakamoto F Y, da Silva M R, et al. Microstructural characterization and mathematical modeling for determination of volume fraction of eutectoid mixture of the cu-8.5wt% sn alloy obtained by unidirectional upward solidification. Mater Sci Forum. 2020;1012 MSF:302–7. Available at: https://doi.org/10.4028/www.scientific.net/MSF.1012.302

[9] Cordero Z C, Knight B E, Schuh C A. Six decades of the Hall–Petch effect – a survey of grain-size strengthening studies on pure metals. *Internacional Materials Reviews* **61** 495, 2016.

[10] Dunstan D J, Bushby A J, Grain size dependence of the strength of metals: The Hall–Petch effect does not scale as the inverse square root of grain size. *International Journal of Plasticity* **53** 56, 2014.

[11] Osorio W R, Garcia A, Modeling dendritic structure and mechanical properties of Zn-Al alloys as a function of solidification conditions. *Materials Science Engineering*. **325** 103, 2002.

[12] Nascimento M S, dos Santos G A, Teram R, dos Santos V T, da Silva M R, Couto A A. Effects of thermal variables of solidification on the microstructure, hardness, and microhardness of Cu-Al-Ni-Fe alloys. Materials (Basel). 2019;12(8). Available at: https://doi.org/10.3390/ma12081267

[13] ASTM E10, Standard Test Method for Brinell Hardness of Metallic Materials. 2012. *ASTM International*.

IOP Conf. Series: Materials Science and Engineering 1222 (2022) 012009 doi:10.1088/1757-899X/1222/1/012009

An Insight into High Entropy Alloys with a Focus on Friction Stir Processing

K Moiduddin[1,a], **MH Abidi**[1,b*], **AN Siddiquee**[2,c], **SH Mian**[1,d] and **MK Mohammed**[1,e]

[1]Advanced Manufacturing Institute, King Saud University, Riyadh-11421, Saudi Arabia

[2]Department of Mechanical Engineering, Faculty of Engineering and Technology, Jamia Millia Islamia, New Delhi-110025, India

[a]khussain1@ksu.edu.sa, [b]mabidi@ksu.edu.sa, [c]ansiddiqui@jmi.ac.in,

[d]smien@ksu.edu.sa, [e]muneerkm@ ksu.edu.sa

Abstract. High entropy alloys (HEAs) have established an unprecedented stronghold in the domain of metallurgy in a relatively short span of time. The incoming era of engineering materials is expected to consist chiefly of these alloys, and therefore the research in this domain is ongoing extensively. Therefore, HEAs have been comprehensively discussed in the present work, covering from the fundamentals of entropy stabilization to the effect of micro-structural modification on specific properties such as wear, creep and fatigue strength. The evolution of HEAs has been elucidated, along with the applications, characterization techniques for their identification and scrutiny. The methods for fabrication of HEAs such as arc melting, magnetron sputtering, laser cladding etc. are included inclusively. Importantly, the role of friction stir processing (FSP) in the advancement of HEAs has been elaborated. FSP induces further heterogeneity in the HEAs, which help to overcome the strength-ductility trade-off. The intricate effect of FSP on micro-structural modification and the consequent effect on mechanical properties has been studied through the light of literature available on such investigations.

1. Introduction

The conventional strengthening mechanisms such as grain boundary, solid solution, heat treatment and cold work etc. induce a proportional sacrifice on ductility which is undesirable. Conventional metals and alloys bear an inverse relationship between the two important properties: Strength and ductility, consequently the strengthening is usually obtained at the cost of ductility and a strength-ductility trade-off is required for high strength at adequate ductility [1, 2]. Heterogeneity in the monolithic materials can accommodate contradicting properties. This has attracted a lot of researchers on designed multi-component and multiphase alloys [3], composites [4] and even bimodality to achieve high strength – ductility combination. The microstructural heterogeneity such as multimodality in grain size, morphology and/or chemical elements introduce heterogeneity. The introduction of particles of different materials into a substrate via solid state processing by Friction Stir Processing (FSP) is a common route which can introduce some kind of heterogeneities [5-7].

High entropy alloys (HEAs) design-based materials which possess multiple-element as nearly equiatomic concentration in single phase solid solution systems with heterogeneity of atomic distribution [8, 9]. The constituent elements may be present in single parent lattices face-centered cubic (FCC) (in FCC HEAs), body-centered cubic (in BCC HEAs) or hexagonal close packed (in HCP HEAs) structures. The HEAs inherit major property associated with the parent lattice. Multi-element systems usually possess brittle intermetallic phases which are detrimental, whereas the solid solution tendency in HEAs retards the formation detrimental IMCs. The exceptional properties possessed by HEAs such as hot strength [10, 11], combination of high strength and ductility in cryogenic conditions [12], excellent fatigue resistance [13], and corrosion resistance [14, 15], through multi-phase or eutectic-structure design. The original HEAs are recent design-based materials about one-an-half decade old [8, 16]. They are designed on the principles that single-phase solid solution of multiple principal elements maximized configurational entropy and result in a massive single solid solution phase [17]. The condition of equi-atomic concentration of principal elements in a single phase has put a very strict restriction on the production of HEAs and it has also limited a vast space which may provide favorable properties in the condition of equi-atomic concentration condition are removed.

Generally, the HEAs are produced via initial liquid phase such as by casting, plasma arc melting and additive manufacturing etc. Of late alloying via solid state routes such as powder metallurgy, milling and have also attracted the attention. The friction-stir processing (FSP) route has attracted a lot of attraction in the homogenization of the cast structure of HEAs. Other severe plastic deformation (SPD) processes such as equal channel angular pressing (ECAP) [18] and high-pressure torsion (HPT) [19] are reported to refine the structure of HEAs which result in significant improvement in mechanical properties and mainly strength. The variation in strain at different radial distances from the tool center [20] coupled with intense stirring during FSP produces alloying effect with extensive heterogeneity in elemental concentration and grain size [21]. The effect of homogeneous solid solution in varying elemental concentration and bimodality distribution has been demonstrated in case of CuNiMgZn based systems. The solid-state routes rely non-equilibrium diffusion under conditions of pressure, temperature and plastic deformation and are generally referred to as mechanical alloying (MA). Unlike regular alloys, the HEAs with heterogeneity show excellent tunability to deformation by adjusting it in the phase variation caused by: chemical concentration, transformation induced plasticity (TRIP) and grain boundary area variations etc. The change in mechanical properties vary due to heterogeneity assisted tunability which is substantially different for BCC [22, 23]/FCC [24] and HCP HEAs [25].

The HEAs are generally produced by melting in vacuum furnace by using pure metals with a nominal composition at% concentrations followed by casting. The cast ingot may be subsequently treated to produce the HEAs in desired TRIP state. A typical treatment may comprise of: (i) hot-rolling (at 900 °C) to 50% thickness of 50%, (ii) homogenization at 1200 C for 5 h in Ar atmosphere followed by (iii) quenching in ice water (Figure 1).

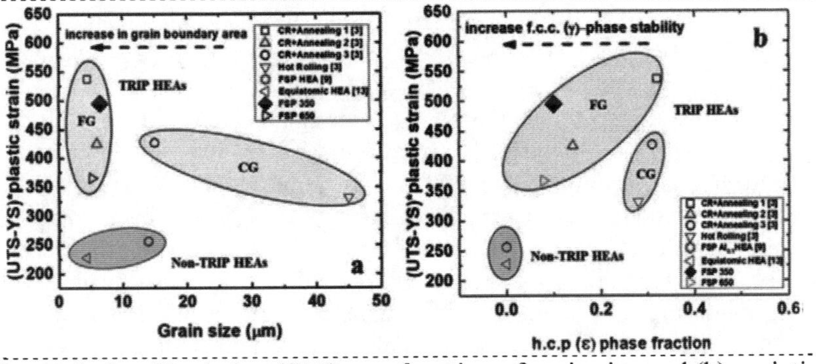

Figure 1. (a) Strength-ductility index (SDI) as a function of grain size and (b) variation of SDI as a function of ε phase fraction (prior to deformation) for grain-refined TRIP HEAs and non-TRIP HEAs. CG: coarse grained; FG: fine grained; CR: cold rolled.

Originally the HEAs are defined as equiatomic alloys with high configurational entropy of mixing (ΔS_{mix}) [> 1.5R (R=8.31 J K^{-1} mol^{-1}, the gas constant)] to overcome the effect of enthalpy [26]. There are common alloys with non-equiatomic high concentration MEAs. Some conventional alloys containing high concentrations of alloying elements such as such as 316 steel (1.15 R), Inconel 718 superalloy (1.31 R), Stellite 6 superalloy (1.13 R). These alloys possess values of (ΔS_{mix}) between R and 1.5R and are known as medium entropy alloys MEAs [27]. The low entropy alloys have the value of (ΔS_{mix}) <R. Unlike solid solution and phase rules for conventional solid solution systems, the phase formation in HEAs are governed not only by configurational entropy of mixing, but also by mismatch entropy, mixing enthalpy, atomic size difference, valency and electronegativity [12, 28]. The solid solution is stable when the difference in atomic size is small and mixing enthalpy is negligible.
Configurational entropy of mixing for a material refers to the distinguishable ways in which the atoms can be arranged in a lattice. The large value of ΔS_{mix} decreases the free energy substantially and stabilizes solid solution phases. Mixing of two different elements placed in physical contact leads to the increase of entropy and which can be ascertained by Boltzmann's hypothesis [29] as per Eqn-(1).

$$\Delta S_{mix} = -R \sum_{i=1}^{n} x_i log x_i \tag{1}$$

where R is the gas constant, n is the number of total elements, x_i is the concentration of element i.
Manufacturing of HEAs by MA has been demonstrated as early as in 2008 [30]. The MA is normally utilized to produce the HEA nanoparticles which are wear resistant and useful in bearing applications. The soft-phase (e.g. Bi) and solid lubricant phase (e.g. MoS_2 etc.) dispersion in HEA matrix such as HEA-Pb, and HEA-Bi system are known to be effective in enhancing wear resistance [31]. The MA is mainly used to synthesize nano-structured HEA in nano-powder form. The HEA nano-powder is subsequently consolidated into bulk HEAs by various techniques but generally spark plasma sintering (SPS). The HEAs are known for slower diffusion but are metastable in nature due to which if SPS not performed under desirable conditions of pressure, heating rate, sintering temperature and cooling rate it will risk phase transformation and the property gain may be lost. A sintering temperature with lower heating rate is advantageous for optimum density and homogenous nanocrystalline microstructure. The sintering pressure of the order of 30-50 MPa is found to provide adequate densification. If the heating temperature and heating rate is not appropriate, a phase transformation during subsequent cooling may be disastrous to the HEA effect.

2. HEAs Evolution and Categories

As shown in Figure 2, one predominant element was combined with other elements in small amounts in the initial stages of alloying to enhance unique properties. Iron (Fe) alloys, copper (Cu) alloys, aluminum (Al) alloys, magnesium (Mg) alloys, titanium (Ti) alloys, and nickel (Ni) alloys are few of the common or essential alloys. Human discovery of metallic materials began with pure metals (e.g., Cu), then progressed to binary (e.g., Fe-Cu, Cu-Ni, etc.), ternary (e.g., Ni-chromium (Cr)-Al), quaternary (e.g., Ni-cobalt (Co)-Al-Cr, Cr-Fe-Co-Ni, Al-Cr-Co-Ni, Al-Fe-Co-Ni, Al-Cr-Fe-Co, etc.), quinary (e.g., Ni-Fe-Cr-Ti-Al, Al-Cr-Fe-Co-Ni, Cr- Manganese (Mn)-Fe-Co-Ni, etc.), and higher-order alloys to achieve favorable properties in order to meet ever-increasing demands [32, 33]. Binary and ternary alloys are the most common types of conventional alloys. A binary alloy is described as a blend of only two types of constituents, while a ternary alloy is made up of three different kinds of elements such as Fe, Ni, and Cr. Ternary alloys of Fe and Ni with Cr and Al are the most important materials for many high temperature applications, especially for gas turbines [34]. Likewise, a quaternary alloy is made up of four constituents, whereas a quinary alloy is made up of five. The complexity of the elemental composition in alloys has been gradually increasing over time as shown in Figure 2 [35, 36].

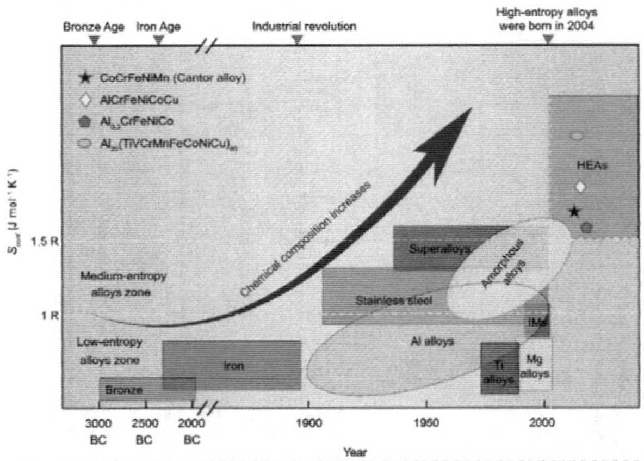

Figure 2. Rising trend of alloy chemical complexity versus time (IMs: intermetallic or metallic compounds, HEA: high-entropy alloy) [35, 36].

HEAs are the latest trend, and they are characterized as materials with complex multi-element compositions. In comparison to traditional alloys, they have a striking functionality and their high structure entropy mixing is more stable at high temperatures. Figure 3 shows the X-ray Powder Diffraction (XRD) trends of a range of binary to septenary alloys, demonstrating the high-entropy effect [37]. In quinary, senary (six constituents), and septenary (seven elements) alloys, the phases are relatively simple: there are only two major phases, both of which have simple structures such as body-centered cubic (BCC) and Face Centered Cubic Structure (FCC). The presence of such basic structures defies perceptions that there might be a development of a number of binary and ternary compounds in HEAs. Generally, FCC-type HEAs are soft and malleable and the BCC-type HEAs have high strength but are usually brittle [37]. Aside from the high entropy effect, another important characteristic for HEAs is their significant lattice distortion, which is caused by a large atomic size discrepancy between the constituent alloying elements.

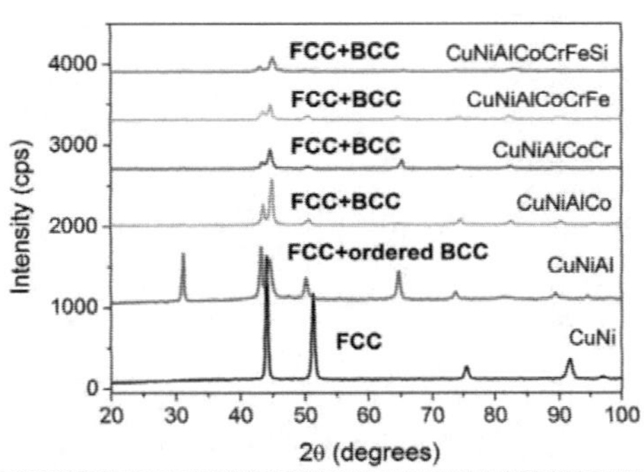

Figure 3. XRD patterns of a series alloy generated by adding one additional element to the preceding one in a sequential order [38].

As an example, all the possible quaternary, ternary, binary, and pure metal subsets of the quinary high-entropy alloy Fe-Ni-Co-Cr-Mn (also known as Cantor alloy) is shown in Figure 4 [39]. Equi-atomic Fe-Ni-Co-Cr-Mn alloy has been reported to display promising strength and ductility at cryogenic temperature as well as it exhibits single-phase with a FCC crystal structure [40].

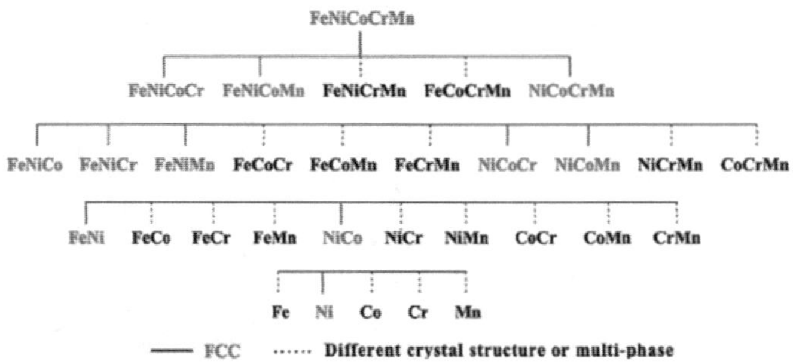

Figure 4. Sequence of high-entropy alloy Fe-Ni-Co-Cr-Mn [39].

3. Properties and applications

HEAs exhibit excellent combinations of strength, ductility, oxidation resistance, and thermal Stability. They have the potential as functional and structural materials for wide range of applications. In conventional alloys the dominant material defines the material properties and other minor alloying elements enhance some of the properties. Whereas in case of HEAs it is the structure type that define and control the material properties. Small alloying elements can also be added to HEAs to increase or decrease a particular property, but it may result in unwanted effects on the other properties [41]. In general BCC-structured HEAs have very high yield strengths and limited plasticity, whereas the FCC structured HEAs have low yield strength and high plasticity. The mixture of both BCC and FCC have shown high yield strength and good ductility. HEAs are mostly more prone to phase transformation under applied temperature or stress, which could be used to trigger interesting plasticity mechanisms as well as to accommodate larger strains. HEAs exhibit high melting points, and sustain high yield strengths even at high temperatures. Senkov et al. [42] produced BCC based $Nb_{25}Mo_{25}Ta_{25}W_{25}$ and $V_{20}Nb_{20}Mo_{20}Ta_{20}W_{20}$, refractory HEAs through vacuum arc-melting. The yield strength of fabricated HEAs were found to much higher as compared to the super alloys especially at higher temperatures as shown in Figure 5 [42]. Refractory metal HEAs find applications in thermal barrier coatings.

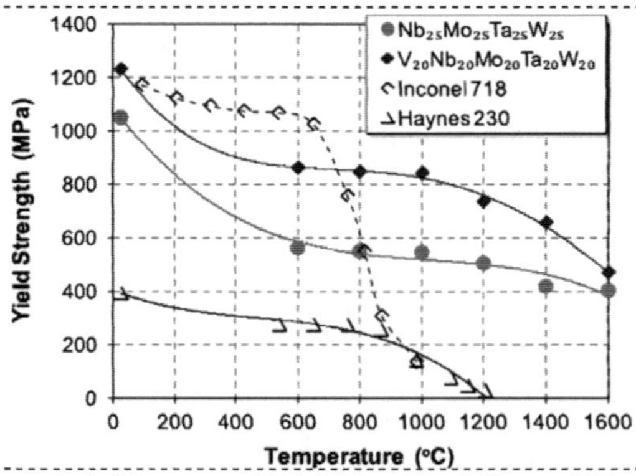

Figure 5. The comparison of yield stress of $Nb_{25}Mo_{25}Ta_{25}W_{25}$ and $V_{20}Nb_{20}Mo_{20}Ta_{20}W_{20}$ HEAs, Inconel 718 and Haynes 230 with respect to temperature [42].

The mechanical properties of HEAs are well within those observed in martensitic steels, advanced high strength steels, nickel-based alloys, etc. However, HEAs with superior properties can be obtained with further optimizations. Several studies have reported HEAs with high hardness and high compressive strength both at room temperature and elevated temperatures [10, 43]. Senkov et al. [43] produced a refractory alloy, $Ta_{20}Nb_{20}Hf_{20}Zr_{20}Ti_{20}$ with predominantly a BCC structure by vacuum arc-melting followed by hot isostatic. The alloy was found to have high compression yield strength and ductility. In addition to the structure types, alloying effect and cooling temperature also effect the mechanical behavior of the HEAs at room temperature. Ma and Zhang [41] studied the effect of Nb addition on the microstructure and properties of AlCoCrFeNi HEAs. The results showed formation of two phases; one BCC solid solution phase and other the Laves phase of (CoCr)Nb type. The compressive yield strength and Vickers hardness were found to increase almost linearly with increasing Nb content. Wang et al. [44] studied the cooling rate and size effect on the microstructure and properties of the AlCoCrFeNi HEAs. Higher cooling rates were linked to uniform microstructures and slight increase in both the strength and the plasticity. Whereas heat treatment induces the grain refinements in the microstructure and influence the mechanical behavior at elevated temperatures. The heat treated HEAs have fine granular structure. They demonstrate homogeneous flow, great resistance to necking and exceptional elongations [45].

Fatigue behavior of the HEAs is comparable with that of conventional alloys and bulk metallic glasses. Some HEAs exhibit fatigue resistance favorable to that of austenitic stainless steels and TWIP steels [46]. Due to the high fatigue resistance they have potential applications in aerospace and automobile industries. In general, HEAs have a fatigue endurance limit of between 540 and 945 MPa and the ratio of fatigue endurance limit to ultimate tensile strength between 0.402 and 0.703. The lower bound of the fatigue ratios are comparable to steel, titanium and nickel alloys and outperforms zirconium alloys. Higher fatigue strength of HEAs is attributed to their higher tensile strength [47].
HEAs have excellent wear resistance properties with potential applications in tools, molds and structural components. Chuang et al. [15] reported the wear resistance of the $Co_{1.5}CrFeNi_{1.5}Ti$ and $Al_{0.2}Co_{1.5}CrFeNi_{1.5}Ti$ alloys to be around two times better than that of conventional wear-resistant steels such as SUJ2 and SKH51. The excellent anti-oxidation property and resistance to thermal softening in these high-entropy alloys were considered to be the main reasons for the outstanding wear resistance. Furthermore, alloying also affects the wear behavior. Wu et al. [48] investigated the wear behavior of AlxCoCrCuFeNi with varying Al content. Higher Al content was found to increase the volume fraction of BCC phase, hardness, and thus the wear resistance. Some HEAs can exhibit ferromagnetic properties [41], and super-paramagnetic properties [49] and hence have potential for electronic applications.

Table 1. Some of research work for properties studies of HEAs.

HEA under consideration	Author	Investigated properties
$Nb_{25}Mo_{25}Ta_{25}W_{25}$ and $V_{20}Nb_{20}Mo_{20}Ta_{20}W_{20}$	Senkov et al. [42]	Mechanical properties of refractory HEAs
TaNbHfZrTi	Senkov et al. [43]	Microstructure and room temperature properties
TaNbHfZrTi	Senkov et al. [10]	Microstructure and elevated temperature properties
AlCoCrFeNi	Wang et al. [44]	Cooling rate and size effect on microstructure and properties
AlCoCrFeNi	Ma and Zhang [41]	Ferromagnetic properties
AlCrCuNiFeCo	Kuznetsov [45]	Tensile properties
CrMnFeCoNi	Thurston [46]	Fatigue resistance
$Al_{0.5}CoCrCuFeNi$	Hemphill et al. [47]	Fatigue behavior
$Co_{1.5}CrFeNi_{1.5}Ti$ and $Al_{0.2}Co_{1.5}CrFeNi_{1.5}Ti$	Chuang et al. [15]	Wear behavior
AlxCoCrCuFeNi	Wu et al. [48]	Wear behavior as a function of aluminum content
CoCrCuFeNiTix	Wang et al. [49]	Super-paramagnetic properties

4. Major characterization techniques employed to identify HEAs, inter-metallic compounds

Microstructure of HEAs vary based on the fabrication technique, heat treatment, and other operations such as rolling or forging. Since microstructure is the key factor which governs the material properties, a detailed characterization of the microstructure is essential in order to interpret the material properties. Moreover, since the phase structure of HEAs strongly influence the performance and applications it is imperative to understand phase structures in order to design HEAs with specific properties. Phases in HEAs are usually solid solution, intermetallic compounds, amorphous alloy or a combination of them. BCC, FCC and sometimes HCP constitute the solid phases. Whereas in case of intermetallic there are various m, c and laves phase. The SEM analysis is generally carried out for the microstructural studies, chemical composition is characterized by the EDS. The phases and their constituents are characterized by XRD and TEM studies. The grain morphology and surface integrity of HEAs are generally characterized using AFM analysis.

Ye et al. [50] produced $Al_xFeCoNiCuCr$ HEA using laser cladding with varying Al content. Microstructure and compositional analysis was performed using SEM. The SEM analysis revealed the presence of both dendritic (DR) and interdendritic (ID) areas as shown in Figure 6. The interdendritic segregation in the HEA produced by laser cladding were found to be smaller than that of casting route. Since the interdendritic segregation causes adverse impacts on the performance of the alloy, HEAs fabricated by laser cladding were proposed to have better performance than the casting ones. XRD and TEM analysis revealed the influence of Al content on the phase composition. Figure 7 shows the XRD peaks with phase compositions for varying Al content in $Al_xFeCoNiCuCr$ HEA [12]. In case of HEAs for high temperature applications it is required to study the time-dependent plastic deformation of materials. Creep test using nanoindentation equipment with a spherical tip is generally incorporated for the characterization. Ma et al. [51], studied the creep behavior of CoCrFeCuNi HEA in as-deposited and 800 K annealed state using nano-indentation. It was found that both the internal structure and external loading sequence strongly affect the creep behavior.

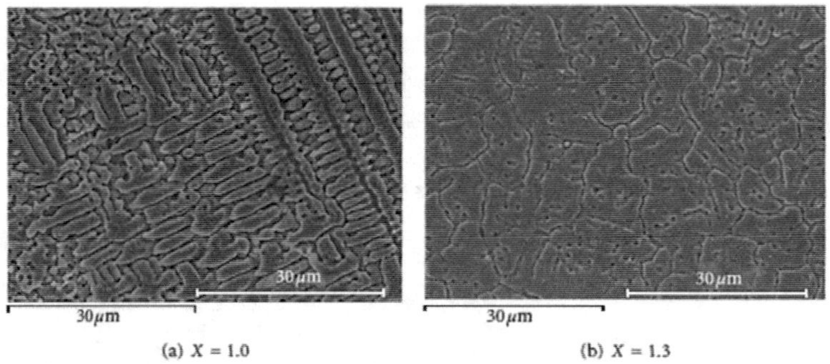

Figure 6. Microstructure of $Al_xFeCoNiCuCr$ HEA with (a) Al 1.0 (b) Al 1.3 [12].

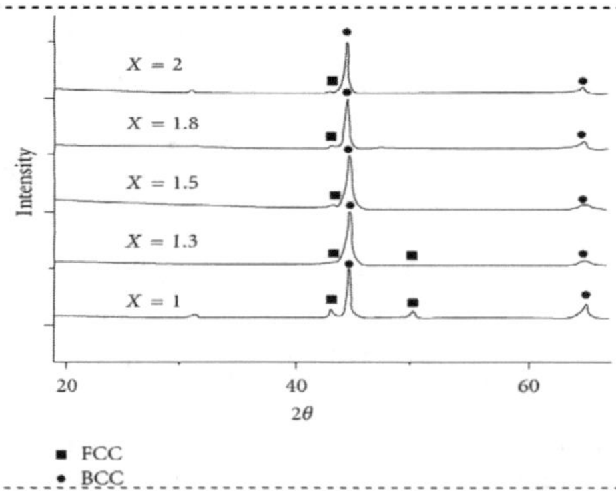

Figure 7. XRD peaks showing the effect of Al content on the phase composition [12].

5. Major routes/methods of manufacturing HEAs

The techniques to produce bulk HEAs from the liquid state involve melting and solidification processes [52]. The solid pure metals are transformed into a liquid state and uniformly blended during the melting process. The liquid mixture is then solidified to develop bulk alloys under various cooling conditions. Another method is to make bulk HEAs from a solid-state, which can be done through powder metallurgy. Furthermore, arc-melting, which has over two hundred documented studies, is the most popular synthesis routes for forming bulk HEAs [53]. HEA coatings can also be manufactured using magnetron sputtering and laser cladding, while small-scale bulk HEAs are produced using splat quenching, laser cladding, and powder metallurgy. The sections that follow outline several of the approaches used in the fabrication or processing of HEAs.

5.1. Arc melting

Vacuum arc melting is widely used to produce alloys with lower melting temperatures. The first and the key method for preparing HEAs is vacuum arc furnace smelting and copper mold casting [54]. The HEAs produced by vacuum arc melting had a clear crystal structure with exceptional properties in terms of microstructure, mechanical properties, and corrosion resistance [55]. Ma et al. [56] used arc melting to make Al0.5CrCuFeNi2 HEAs. Clear FCC solid solution with a dendrite microstructure was observed in the as-cast sample. Senkov et al. [42] developed the refractory NbCrMo0.5Ta0.5TiZr, Nb25Mo25Ta25W25, and V20Nb20Mo20Ta20W20 alloys [57, 58]. The single-phase BCC structure was present in both Nb25Mo25Ta25W25 and V20Nb20Mo20Ta20W20 alloys. Zhang et al. [59] also used an arc melting process to form CrFeVTa0.2W0.2 and CrFeVTa0.1W0.1 alloy, which had superior heat-softening tolerance.

5.2. Bridgman solidification

The Bridgman method is also used to make single-crystal ingots or boules. The method entails heating a material well above the melting point in a container and then cooling it gradually from one side, where a seed crystal is positioned. All along the length of the container, single-crystal material is gradually shaped. The procedure may be performed in either a horizontal or vertical configuration. In a high-pressure furnace, the Bridgman process melts the alloy to the desired composition. There are three temperature zones in the Bridgman furnace. Temperatures above the melting point define the upper region, while the lower zone is cooler than the melting point. As a buffer between the two, an adiabatic zone is used.

Bridgman solidification technology can monitor the growth direction and rate of alloy crystal during the solidification by changing the heat flow direction, temperature gradient, and pulling speed of the solid-liquid interface, resulting in improved alloy structure and properties. Ma et al. [60] effectively prepared single-crystal CoCrFeNiAl0.3 and CoCrFeNiAl HEAs using the Bridgman solidification method. By using the same technique, Zuo et al. [61] also generated FeCoNiAlSi HEAs. As compared to samples produced by arc melting, the alloy's soft magnetic properties were enhanced in the Bridgman solidification approach.

5.3. High gravity combustion synthesis

Li et al. [62] proposed another technique for preparing HEAs using combustion synthesis in a high-gravity setting. The crucible was placed on a Ni-based superalloy rotor with two internal cavities: one for carrying the sample and the other for balancing the opposite sample in this process. The void between the graphite crucible and the cavity wall inside the rotor disk was filled with heat-insulating materials. Once the centrifuge had given a centrifugal force in the combustion chambers greater than 1500g, the chamber device was discharged to a vacuum of about 100 Pa and then ignited with the W wire. The reactive mixtures melted due to the reaction heat and high gravity, and the metal melts were segregated due to the density differential. Eventually, glass ceramics and bulk metal ingots were collected. After that, the metals were machined and polished in preparation for future experiments. CoCrFeNi and AlCoCrFeNi HEAs was produced by high-gravity combustion synthesis.

5.4. Magnetron sputtering

Magnetron sputtering involves bombarding a material's surface with a high-energy particle [55]. Argon can ionize argon-ion and electrons when exposed to a strong electric field. Then, in an electric field, ions are propelled toward the cathode, bombarding the target surface with high energy, causing sputtering. When placed in a magnetic field and an electric field with orthogonal distribution, the target material releases secondary electrons. As a result, the target material ionizes more ions, allowing the bombardment to intensify.

5.5. Laser cladding

Zhang et al. [63] documented a low-cost HEA coating made by laser cladding with a nominal composition of 6FeNiCoCrAlTiSi. An energy source is used to melt the feedstock until it is added to a substrate [64]. The heat source is a directed laser beam, which melts the substrate upon which the feedstock is applied. This technique usually produces a metallurgical bond with higher bond strength. The result is a thick coating with no gaps or porosity. One of the benefits of the laser cladding process is that the laser beam can be directed and localized to a very small region, resulting in a minimal heat-affected zone on the substrate. Furthermore, the lower overall heat decreases the dilution of the coating with substrate materials. The laser-cladding produced coating has a simple BCC solid solution with high micro-hardness, higher softening resistance, and elevated electrical resistivity [65].

5.6. Additive Manufacturing

In addition to the intensive processing requirements of the aforementioned methods, it could be argued that these techniques restrict the shape and size of HEA components. The new development is an additive manufacturing (AM), which can manufacture complex near-net shape components. Kunce et al. developed bulk TiZrNbMoV [66] and ZrTiVCrFeNi [67] alloys through AM. Direct laser manufacturing (DLF) was used by Joseph et al. [53] to manufacture bulk AlxCoCrFeNi HEAs. The feedstock or the raw material in AM is usually pre-alloyed HEA wires or HEA powders. High-energy sources, such as lasers or electron beams, fuse the feedstock. These power sources can reach high temperatures to melt refractory elements quickly and helps reduce the evaporation of the elements with lower melting points. AM technologies such as laser metal deposition (LMD), selective laser melting (SLM), and electron beam melting (EBM) are widely used to fabricate HEAs. LMD was used by Xiang et al. [68] to prepare CrMnFeCoNi HEA and Zhu et al. [69] prepared the same alloy using SLM and discovered that the mechanical properties of the HEA were improved. The additively produced HEA exhibits a remarkable combination of high strength and excellent ductility.

5.7. Friction Stir Processing

Friction stir processing (FSP) is perhaps the most commercially viable thermo-mechanical processing technique for bulk materials, and it can be used for solid-state joining. FSP is a simple, clean, and eco-friendly solid-state processing technique [70]. Figure 8 depicts a diagram of the FSP operation. FSP is a solid-state process in which a specially built rotating cylindrical tool composed of a shoulder with an attached pin is plunged into the processed plate. Friction produces heat at the interface between the shoulder and the processed workpiece, softening the material. The material is stirred by the spinning pin, and the process is completed by the tool moving over the sample. At a high temperature, the material undergoes extreme plastic deformation during the process.

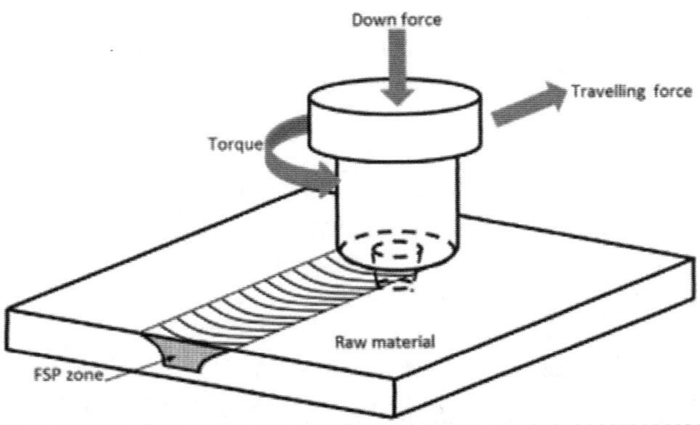

Figure 8. Schematic of friction stir processing process [71].

FSP can be used to improve HEA's mechanical and microstructural properties even further. For example, Li et al. [70] made a novel Al matrix composite reinforced with Al0.8CoCrFeNi HEA particles, and Gao et al. [72] used a multi-pass FSP to fabricate Fe-Co-Ni-Cr-Al HEA reinforced Al5083 composites. The XRD outcomes revealed that FSP technology can be used to make HEA reinforced Al matrix composites that are free of intermetallics. The composites manufactured from FSP had greater microhardness, yield strength, ultimate tensile strength, and wear resistance than the base alloy and FSPed samples without the HEA particles, according to the findings. Besides, Kumar et al. [20] used FSP to manipulate the microstructure of a HEA. This study aimed to see how microstructural refinement using FSP affected the plastic deformation behavior of a HEA. The HEA displayed an excellent combination of strength and ductility in the FSP condition, as well as high values of friction stress and Hall–Petch coefficient, suggesting that grain refinement can be a very powerful HEA strengthening mechanism. The FSP-engineered microstructure of the transformation-induced plasticity (TRIP) assisted HEA with composition $Fe_{50}Mn_{30}Co_{10}Cr_{10}$ (at. %) had a significantly smaller grain size, and optimized fractions of FCC and hexagonal close-packed (hcp) phases [25]. The effect of friction stir welding (FSW) parameters on the structure and properties of $Fe_{49}Mn_{30}Cr_{10}Co_{10}C_1$ HEA welds was investigated by Shaysultanov et al. [73]. FSW of the alloy resulted in recrystallization of the stir zone and a reduction in the fraction of the carbides to 1–2%; however, the percentage of the hcp phase remained virtually unchanged from the initial state. Yang et al. [74] used cooling-assisted FSP to manufacture HEA-reinforced aluminum alloy surface composites. The impact of HEA particle volume fraction on microstructure, microhardness, and tribological properties was investigated. Hardness and wear resistance strengthened as the volume fraction of HEA particles improved. Nair et al. [75] conducted the FSP of an Al0.1CoCrFeNi HEA under monitored cooling conditions. Because of the substantial grain refinement, the hardness of processed HEA was significantly higher than that of as-cast HEA. The processed HEA's improved erosion resistance is linked to their increased hardness, plasticity resistance, and yield strength than the as-cast HEA. The prospect of utilizing FSW as a welding technique to achieve strong joints in HEAs was explored by Zhu et al. [76, 77]. Using this method, the authors were able to weld Co16Cr28Fe28Ni28 and CoCrFeNiAl0.3 HEAs. Finally, grain refinement, progressive recrystallization, hardness improvement, ductility reduction, and material strengthening are some of the advantages that can be associated with the FSP of HEAs [78].

6. Conclusions

The literature encompassing friction stir processing of HEAs has been expounded in the present work, along with a rudimentary perspective of these modern-day engineering materials. Various techniques of manufacturing the HEAs are discussed with their methodology and principles of fabrication. The review suggests that more innovative technologies are likely to evolve for their production, wherein avoiding the formation of detrimental intermetallic compounds is among the primary concerns.

Friction stir processing tremendously enhances the micro-structural and mechanical aspects of these alloys. HEAs have the potential to replace superalloys in the aerospace sectors, and are also excellent refractory materials. Modern characterization techniques such as atomic force microscopy and nano-indentation are vital to the research and development of these alloys.

Acknowledgments
Authors would like to thanks the National Plan for Science, Technology, and Innovation (MAARIFAH), King Abdulaziz City for Science and Technology, Saudi Arabia, for funding this work under Award 3-17-04-001-0003.

References
[1] Koch C et al 1999 Ductility of nanostructured materials *Materials Research Bulletin* **24** p 54-58
[2] Zhu Y T and Liao X 2004 Nanostructured metals retaining ductility *Nature materials* **3** p 351
[3] Kumar K, Van Swygenhoven H and Suresh S 2003 Mechanical behavior of nanocrystalline metals and alloys *Acta Materialia* **51** p 5743-5774
[4] Wegst U G et al 2015 Bioinspired structural materials *Nature materials* p 14-23
[5] Shafiei-Zarghani A, Kashani-Bozorg S and Zarei-Hanzaki A 2009 Microstructures and mechanical properties of Al/Al2O3 surface nano-composite layer produced by friction stir processing *Materials Science and Engineering A* **500** p 84-91
[6] Hsu C, Kao P and Ho N 2005 Ultrafine-grained Al–Al2Cu composite produced in situ by friction stir processing *Scripta Materialia* **53** p 341-345
[7] Wu X and Zhu Y 2017 Heterogeneous materials a new class of materials with unprecedented mechanical properties *Materials Research Letters* **5** p 527 -532
[8] Yeh J et al 2004A Nanostructured high-entropy alloys with multiple principal elements novel alloy design concepts and outcomes *Advanced Engineering Materials* **6** p 299-303
[9] Yeh J et al 2004 Formation of simple crystal structures in Cu-Co-Ni-Cr-Al-Fe-Ti-V alloys with multiprincipal metallic elements *Metallurgical and Materials Transactions A* **35** p 2533-2536
[10] Senkov O N et al 2012 Microstructure and elevated temperature properties of a refractory TaNbHfZrTi alloy *Journal of Materials Science & Technology* **47**(9) p 4062–4074
[11] Chen S T et al 2010 Microstructure and properties of age-hardenable AlxCrFe1 5MnNi0 5 alloys *Materials Science and Engineering A -Structural* **527**(21-22) p 5818–5825
[12] Otto F et al 2013 Relative effects of enthalpy and entropy on the phase stability of equiatomic high-entropy alloys *Acta Materialia* **61** p 2628–2638
[13] Tang Z et al 2015 Fatigue behavior of a wrought Al0 5CoCrCuFeNi two-phase high-entropy alloy *Acta Materialia* **99** p 247–258
[14] Shi Y Z, Yang B and Liaw P K 2017 Corrosion-resistant high-entropy alloys a review *Metals* **7**(2) p 43
[15] Chuang M H et al 2011 Microstructure and wear behavior of AlxCo1 5CrFeNi1 5Tiy high-entropy alloys *Acta Materialia* **59**(16) p 6308–6317
[16] Cantor B et al 2004 Microstructural development in equiatomic multicomponent alloys *Materials Science and Engineering: A* **375-377** p 213–218
[17] Li Z and Raabe D 2017 Strong and Ductile Non-equiatomic High-Entropy Alloys Design Processing Microstructure and Mechanical Properties *JOM - The Journal of The Minerals, Metals & Materials Society (TMS)* **69** p 2099–2106
[18] Heczel A et al (2017 Defect structure and hardness in nanocrystalline CoCrFeMnNi High-Entropy Alloy processed by High-Pressure Torsion *Journal of Alloys and Compounds* **711** p 143–154
[19] Hammond V H et al 2014 Equal channel angular extrusion of a low-density high entropy alloy produced by high energy cryogenic mechanical alloying *JOM - The Journal of The Minerals, Metals & Materials Society (TMS)* **66** p 2021–2029
[20] Kumar N et al 2015 Friction stir processing of a high entropy alloy Al01CoCrFeNi *JOM - The Journal of The Minerals, Metals & Materials Society (TMS)* **67** p 1007–1013

[21] Khan T, Bajaj D and Siddiquee AN 2021 Friction Stir Engineering for fabrication of Ultra-refined CuNiMgZn alloys *Materials Letters* p 129596
[22] Zou Y, Ma H and Spolenak R 2015 Ultrastrong ductile and stable high entropy alloys at small scales *Nature Communications* **6** p 7748
[23] Ma E and Wu X 2019 Tailoring heterogeneities in high-entropy alloys to promote strength-ductility synergy *Nature Communications* **10** p 5623
[24] Wu Z et al 2014 Temperature dependence of the mechanical properties of equiatomic solid solution alloys with face-centered cubic crystal structures *Acta Materialia* **81** p 428–441
[25] Nene S S et al 2017 Enhanced strength and ductility in a friction stir processing engineered dual phase high entropy alloy *Scientific Reports* 7
[26] Zhang Y et al 2014 Microstructures and properties of high-entropy alloys *Progress in Materials Science* **61** p 1-93
[27] Zhou Y et al 2018 Design of non-equiatomic medium entropy alloys *Scientific Reports* p 81236
[28] Guo S et al 2013 More than entropy in high-entropy alloys forming solid solutions or amorphous phase *Intermetallics* **41** p 96–103
[29] Harpole K W and Sharp K A 2011 Calculation of Configurational Entropy with a Boltzmann Quasi-harmonic Model: The Origin of High-Affinity Protein Ligand Binding *The Journal of Physical Chemistry B - ACS Publications* **115** p 9461–9472
[30] Varalakshmi S, Kamaraj M and Murty B S 2008 Synthesis and characterization of nanocrystalline AlFeTiCrZnCu high entropy solid solution by mechanical alloying *Journal of Alloys and Compounds* **460** p 253
[31] Bowden F P and Tabor D *The Friction and Lubrication of Solids* 2001 Oxford university press
[32] Li W et al 2021 Mechanical behavior of high-entropy alloys *Progress in Materials Science* **118** p 100777
[33] Hayun S et al 2020 Enthalpies of formation of high entropy and multicomponent alloys using oxide melt solution calorimetry *Intermetallics* **125** p 106897
[34] Khanna A S *5 - High Temperature Oxidation* in *Handbook of Environmental Degradation of Materials (Second Edition)* M Kutz Editor 2012 William Andrew Publishing Oxford p 127-194
[35] Zhang W, Liaw P K and Zhang Y 2018 Science and technology in high-entropy alloys *Science China Materials* **61**(1) p 2-22
[36] He Q F et al 2017 Design of High-Entropy Alloy A Perspective from Nonideal Mixing *JOM - The Journal of The Minerals, Metals & Materials Society (TMS)* **69**(11) p 2092-2098
[37] Sheikh S A 2018 Alloy Design for High-Entropy Alloys Predicting Solid Solubility and Balancing Mechanical Properties and Oxidation Resistance in *Department of Industrial and Materials Science* Chalmers University of Technology Göteborg Sweden p 96
[38] Tsai M-H and Yeh J-W 2014 High-Entropy Alloys A Critical Review *Materials Research Letters* **2**(3) p 107-123
[39] Wu Z et al 2014 Recovery recrystallization grain growth and phase stability of a family of FCC-structured multi-component equiatomic solid solution alloys *Intermetallics* **46** p 131-140
[40] Wu Z, Parish C M and Bei H 2015 Nano-twin mediated plasticity in carbon-containing FeNiCoCrMn high entropy alloys *Journal of Alloys and Compounds* **647** p 815-822
[41] Ma S G and Zhang Y 2012 Effect of Nb addition on the microstructure and properties of AlCoCrFeNi high-entropy alloy *Materials Science and Engineering A* **532** p 480-486
[42] Senkov O N et al 2011 Mechanical properties of Nb25Mo25Ta25W25 and V20Nb20Mo20Ta20W20 refractory high entropy alloys *Intermetallics* **19**(5) p 698-706
[43] Senkov O N et al 2011 Microstructure and room temperature properties of a high-entropy TaNbHfZrTi alloy *Journal of Alloys and Compounds* **509**(20) p 6043-6048
[44] Wang F J et al 2009 Cooling Rate and Size Effect on the Microstructure and Mechanical Properties of AlCoCrFeNi High Entropy Alloy *Journal of Engineering Materials and Technology* **131**(034501)

[45] Kuznetsov A V et al 2012 Tensile properties of an AlCrCuNiFeCo high-entropy alloy in as-cast and wrought conditions *Materials Science and Engineering A* **533** p 107-118
[46] Thurston K V S et al 2017 Effect of temperature on the fatigue-crack growth behavior of the high-entropy alloy CrMnFeCoNi *Intermetallics* **88** p 65-72
[47] Hemphill M A et al 2012 Fatigue behavior of Al05CoCrCuFeNi high entropy alloys *Acta Materialia* **60**(16) p 5723-5734
[48] Wu J-M et al 2006 Adhesive wear behavior of AlxCoCrCuFeNi high-entropy alloys as a function of aluminum content *Wear* **261**(5) p 513-519
[49] Wang X F et al 2007 Novel microstructure and properties of multicomponent CoCrCuFeNiTix alloys *Intermetallics* **15**(3) p 357-362
[50] Ye X et al 2011 Synthesis and Characterization of High-Entropy Alloy A l *Advances in Materials Science and Engineering* 2011 p e485942
[51] Ma Y et al 2015 Nanoindentation creep behavior in a CoCrFeCuNi high-entropy alloy film with two different structure states *Materials Science and Engineering A* **621** p 111-117
[52] Zhang Y and Xing Q *High Entropy Alloys Manufacturing Routes* in *Encyclopedia of Materials Metals and Alloys* 2020
[53] Joseph J et al 2015 Comparative study of the microstructures and mechanical properties of direct laser fabricated and arc-melted AlxCoCrFeNi high entropy alloys *Materials Science and Engineering A* **633** p 184-193
[54] Zhang Y *Preparation Methods of High-Entropy Materials* in *High-Entropy Materials An Introduction* Y Zhang Editor 2019 Springer Nature Singapore p 65-75
[55] Yin X and S Xu 2018 Properties and Preparation of High Entropy Alloys *MATEC Web of Conferences* **142** p 03003
[56] Ma S G et al 2015 Microstructural features and tensile behaviors of the Al05CrCuFeNi2 high-entropy alloys by cold rolling and subsequent annealing *Materials & Design* **88** p 1057-1062
[57] Senkov O N et al 2010 Refractory high-entropy alloys *Intermetallics* **18**(9) p 1758-1765
[58] Senkov O N and CF Woodward 2011 Microstructure and properties of a refractory NbCrMo05Ta05TiZr alloy *Materials Science and Engineering A* **529** p 311-320
[59] Zhang W, Liaw P K and Zhang Y 2018 A Novel Low-Activation VCrFeTaxWx (x = 01 02 03 04 and 1) High-Entropy Alloys with Excellent Heat-Softening Resistance *Entropy* **20**(12) p 951
[60] Ma S G et al 2013 A Successful Synthesis of the CoCrFeNiAl03 Single-Crystal High-Entropy Alloy by Bridgman Solidification *JOM - The Journal of The Minerals, Metals & Materials Society (TMS)* **65**(12) p 1751-1758
[61] Zuo T et al 2015 Influence of Bridgman solidification on microstructures and magnetic behaviors of a non-equiatomic FeCoNiAlSi high-entropy alloy *Intermetallics* **67** p 171-176
[62] Li R X, Liaw P K and Zhang Y 2017 Synthesis of AlxCoCrFeNi high-entropy alloys by high-gravity combustion from oxides *Materials Science and Engineering A* **707** p 668-673
[63] Zhang H, Pan Y, and He Y 2011 Effects of Annealing on the Microstructure and Properties of 6FeNiCoCrAlTiSi High-Entropy Alloy Coating Prepared by Laser Cladding *Journal of Thermal Spray Technology* **20**(5) p 1049-1055
[64] Zhang Y et al 2014 Microstructures and properties of high-entropy alloys *Progress in Materials Science* **61** p 1-93
[65] Man H C et al 2001 Microstructure and formation mechanism of in situ synthesized TiC/Ti surface MMC on Ti6Al4V by laser cladding *Scripta Materialia - SCRIPTA MATER* **44** p 2801-2807
[66] Kunce I, Polanski M and Bystrzycki j 2014 Microstructure and hydrogen storage properties of a TiZrNbMoV high entropy alloy synthesized using Laser Engineered Net Shaping (LENS) *International Journal of Hydrogen Energy* **39**(18) p 9904-9910
[67] Kunce I, Polanski M and Bystrzycki J2013 Structure and hydrogen storage properties of a high entropy ZrTiVCrFeNi alloy synthesized using Laser Engineered Net Shaping (LENS) *International Journal of Hydrogen Energy* **38**(27) p 12180-12189

[68] Xiang S et al 2019 Microstructures and mechanical properties of CrMnFeCoNi high entropy alloys fabricated using laser metal deposition technique *Journal of Alloys and Compounds* **773** p 387-392
[69] Zhu Z G et al 2018 Hierarchical microstructure and strengthening mechanisms of a CoCrFeNiMn high entropy alloy additively manufactured by selective laser melting *Scripta Materialia* **154** p 20-24
[70] Li J et al 2020 Friction stir processing of high-entropy alloy reinforced aluminum matrix composites for mechanical properties enhancement *Materials Science and Engineering A* **792** p 139755
[71] Weglowski M, Dymek S and Hamilton C 2013 Experimental investigation and modelling of Friction Stir Processing of cast aluminium alloy AlSi9Mg *Bulletin of the Polish Academy of Sciences Technical Sciences* **61** p 893-904
[72] Gao J et al 2020 Producing of FeCoNiCrAl high-entropy alloy reinforced Al composites via friction stir processing technology *The International Journal of Advanced Manufacturing Technology* **110**(1) p 569-580
[73] Shaysultanov D et al 2021 Friction Stir Welding of a TRIP Fe49Mn30Cr10Co10C1 High Entropy Alloy *Metals* **11**(1)
[74] Yang X et al 2020 Microstructural Microhardness and tribological analysis of cooling-assisted friction stir processing of high-entropy alloy particles reinforced aluminum alloy surface composites *Surface Topography Metrology and Properties* **8**(3) p 035012
[75] Nair R B, Arora H S and Grewal H S 2020 Enhanced cavitation erosion resistance of a friction stir processed high entropy alloy *International Journal of Minerals Metallurgy and Materials* **27**(10) p 1353-1362
[76] Zhu Z G et al 2017 Friction stir welding of a CoCrFeNiAl03 high entropy alloy *Materials Letters* **205** p 142-144
[77] Zhu Z G et al 2018 Friction-stir welding of a ductile high entropy alloy microstructural evolution and weld strength *Materials Science and Engineering A* **711** p 524-532
[78] Garcia Filho F C and Monteiro S N 2020 Welding Joints in High Entropy Alloys A Short-Review on Recent Trends *Materials* **13**(6)

Effect of norcholine fluid derivative on the resilience of mild steel to electrochemical degradation in weak HCl media

R T Loto
Department of Mechanical Engineering, Covenant University, Ogun state, Nigeria

tolulope.loto@covenantuniversity.edu.ng

Abstract. The resilience of mild steel to electrochemical degeneration in weak electrolyte corrosion (0.5 M HCl) was assessed at discrete concentrations of Norcholine derivatives by potentiodynamic polarization and open circuit potential analysis. Degradation rate of the steel substantially declined with augmentation of Norcholine concentration from 9.49 mm/y in control HCl (no Norcholine) to 0.27 mm/y at highest Norcholine value of 15%. The degradation rate at this inhibitor concentration correlates of 97.1% protection performance. Norcholine demonstrated cathodic inhibition effect from observation of the negative deviation in corrosion potential. Norcholine was also observed to indicate substantial passive reaction mechanism on the anodic polarization regions of the polarization graphs. This translates to evolution of non-porous layer on the steel exterior. Open circuit potential graphs indicate substantial variation between the inhibited and acid degraded graphs of the steel. Mathematical computations through analysis of variance demonstrates Norcholine concentration and potential tracking time are statistically important variation sources influencing the protection performance of Norcholine at values of 68.03% and 26.4%.

1. Introduction

Carbon steel exhibits universal utilization in most manufacturing and production plants due to its availability, cost, recyclability, ease of production and fabrication and being the most widely produced steel globally. However, the steel corrodes easily in mild aqueous environments. This is having been a major problem for scientist, engineers and industrialist sue to lack of comprehensive cost-effective replacements [1, 2]. The presence of corrosive anions such as SO_4^{2-}, Cl^- etc. in industrial operating conditions is a major problem for carbon steels limiting their operational lifespan. This is mostly common in industries where acidic solution is used various applications such as raw materials for the production of finished goods, intermediate compounds, acid production and industrial operations such as oil-well acidizing, acid descaling etc. [3-5]. HCl is extensively applied in dye production, manufacture of batteries, photoflash bulbs and in crude distillation overhead system. The low cost of carbon steel coupled with its extensive application necessitates the need for low cost corrosion prevention method. Chemical derivative known as corrosion inhibitors has seen applications universally in the protection of metallic alloys from the reaction effect of corrosive species. The harmful nature of most compounds also necessitates the need for sustainable compounds for corrosion inhibition [6]. Compounds composed molecules such as O, P, N and S etc. tends to adequately suppressing material degradation [7–9]. Organic compounds generally prevent corrosion reaction processes through adsorption mechanisms which inhibit the electrolytic transport and degradation reaction of anions on the steel [10-17]. The structural dimension and adsorption mechanism of the compound are also important in determining effective corrosion inhibiting organic compounds [18-22]. This article studies the corrosion suppression properties of Norcholine on the degradation of low carbon steel in dilute HCl solution.

2. Experimental methods

Mild steel (MS) rod bought in Lagos, Nigeria has a diametric dimension of 14.5 mm. The steel was machined into 7 test exhibits with minimal length of 6 mm and cleansed with distilled H_2O and C_3H_6O. Norcholine (NHL), a light-yellow liquid was procured from Sigma Aldrich company in the United States. The fluid derivative has a molecular formular and molar mass of $C_4H_{11}NO$ and 89.14 g mol^{-1}. NHL was prepared in volumetric concentration from 2.5% to 15% in 200 mL of 0.5 M HCl solution. Analar grade NaCl was added at 3.5% concentration per 200 mL of acid/inhibitor solution. Potentiodynamic polarization test was done using a ternary terminal configuration (Pt counter terminal, MS specimen terminal and Ag/AgCl standard terminal) immersed in the acid-inhibitor solution within a specialized container, and linked to Autolab PGSTAT 30 potentiostatic device and monitoring computer for real-time plotting. The plots were recorded from -1.5V to +1.5 mV at scanning rate of 0.0015V/s. Open circuit potential analysis was performed using the same potentiostat with two terminal configurations (MS specimen terminal and Ag/AgCl standard terminal) linked to a computer and monitored for 288 h at 48 h interval. 2-factor single level statistical test (ANOVA F - test) was employed to assess the statistical relevance of NHL concentration and potential tracking time (independent variables) on the inhibition performance of Norcholine at confidence level of 95% (significance level of $\alpha = 0.05$).

3. Results and discussion

3.1 Potentiodynamic polarization studies

Graphical representations depicting the electrochemical behaviour of MS in 0.5 M HCl at specific NHL solution is shown in Fig. 1 Table 1 depicts the outputs from the graphs. Corrosion rate of MS at 0% NHL is observed to be 9.47 mm/y due to the redox reaction mechanisms induced by chloride species. The chloride caused the anodic degradation of MS surface resulting in accelerated degeneration of the steel. Addition of NHL at 2.5% concentration significantly decreased the degradation rate of MS to 3.85 mm/y which corresponds to protection performance of 59.35%. At 15% NHL concentration the degradation rate is 0.27 mm/y correlating to protection performance of 97.1%. More augmentation of NHL concentration aided continuous decrease in corrosion rate and substantial rise in protection performance, till 0.27 mm/y and 97.10% at 15% NHL concentration. Decrease in degradation rate analogous to rise in NHL concentration is because of increase in NHL molecules which suppressed the chloride acceleration oxidation-reduction reaction processes. The corrosion potential transited significantly in the cathodic direction relative to the potential of MS at 0% NHL beyond the threshold 85mV signifying NHL is a cathodic inhibitor. Secondly it shows NH compound suppresses corrosion of MS through selective precipitation on active sites and modification of the corrosive medium. The cathodic portion of the polarization plot from 7.5% NHl concentration shows the cathodic inhibition process occurs under activation control. Simultaneously, the anodic portion of the polarization plot shows significant passivation behaviour due to stifling of the corrosion reaction mechanism and accumulation of protonated NHL molecules on MS surface.

Table 1. Data for potentiodynamic polarization of MS in 0.5 M HCl solution at specific NHL concentration

Inhibitor Conc. (%)	Inhibitor Conc. (M)	Corrosion Rate (mm/y)	Inhibition Efficiency (%)	Corrosion Current density (A/cm²)	Corrosion Potential (V)	Polarization Resistance, R_p (Ω)	Cathodic Tafel Slope, B_c (V/dec)	Anodic Tafel Slope, B_a (V/dec)
0	0	9.49	0	8.31E-04	-0.360	25.99	-0.102	0.419
2.5	2.80E-04	3.85	59.35	3.37E-04	-0.429	53.79	-0.090	0.292
5	5.60E-04	3.16	66.61	2.77E-04	-0.763	116.90	-0.251	0.374
7.5	8.40E-04	2.61	72.41	2.29E-04	-0.853	188.83	-0.102	0.325
10	1.12E-03	1.86	80.40	1.63E-04	-0.859	298.36	-0.192	0.335
12.5	1.40E-03	0.55	94.19	4.82E-05	-0.858	403.78	-0.147	0.290
15	1.68E-03	0.27	97.10	2.41E-05	-0.872	547.61	-0.059	0.105

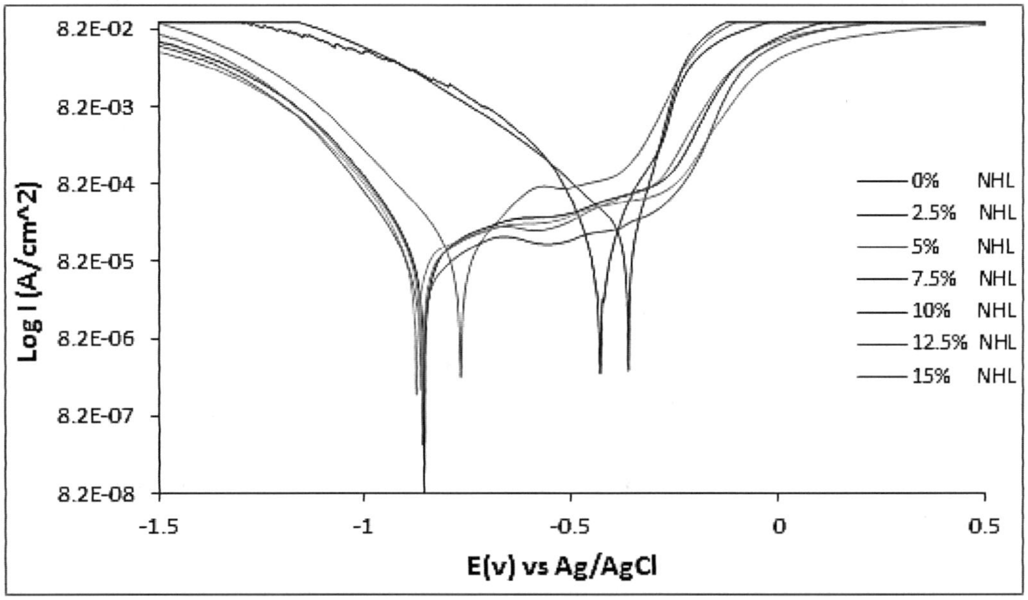

Figure. 1 Graphical representation of the potentiodynamic polarization behavior of in 0.5 M HCl solution at specific NHL concentration

3.2 Open circuit potential analysis

The open-circuit potential plots MS corrosion is shown in Fig. 2. General equilibrium potential of about was attained at 1 h of exposure in HCl solution at all NHL concentrations. Anodic potential shift occurred with increase in NHL concentration. This is because of the presence of adequate ionized NHL molecules which adsorbed unto MS surface forming chemical complexes that stifled further interfacial reaction and chloride adsorption detrimental to MS. MS attained mean steady state corrosion potential value ~306 mV at 10% NHL concentration. The potential plot slightly differs with respect to NHL concentration. At 2.5% - 5% NHL concentration, the thermodynamic tendency to corrode is high. At 7.5% NHL concentration, the tendency to corrode is intermediate while at 10% - 15%, the corrosion tendency is very low signifying strong dependence of NHL performance on concentration. The corrosion potential plot of MS at 0% NHL was significantly cathodic throughout due to active deterioration of the steel surface by chloride anions. Generally, the plots at specific NHL concentration were thermodynamically stable while the anodic shift with respect to NHL concentration signifies enhancement of the protective properties NHL compound due to molecular adehesion of NHL cations unto the steel exterior and maintains the potential values in the passive potential region of the plot.

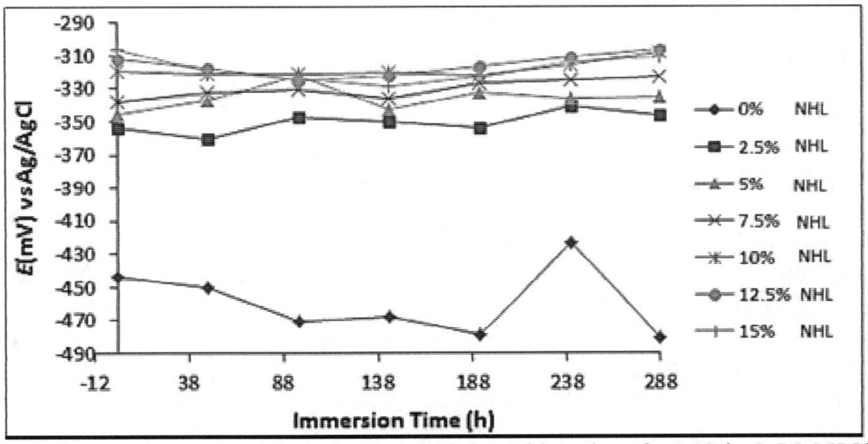

Figure. 2 Plot of corrosion potential against potential tracking time for MS in 0.5 M HCl at specific NHL concentrations

3.3 Statistical analysis

ANOVA data indicating the mathematical relevance of exposure time and NHL concentration on the corrosion protection effect of NHL shows both variables influence the performance of NHL at different levels. However, the percentage importance of NHL concentration stands at 68.03% compared to the potential tracking time which stands at 26.4%. The F-values depicted at 28.28 and 11.70 are significantly more than the significance factor ($\alpha = 0.05$). The results expresses NHL concentration and observation time are significant independent variables which affects the protection effect of NHL on MS.

Table 2. Statistical analysis (ANOVA) for NHL inhibition performance on MS at 95% confidence level

Source of Variation	Mean Square Ratio	Significance F	F (%)
NHL Concentration	28.28	2.71	68.03
Exposure Time	11.70	2.87	26.4

4. Conclusion

Norcholine effectively inhibited mild steel degeneration in dilute HCl solution with respect to its concentration. The compound exhibited cathodic inhibition properties. Norcholine initiated passivation reaction processes on the polarization plots of mild steel at specific concentration of norcholine. In the absence of Norcholine, open circuit potential graphs for mild steel reaction were substantially more electronegative than the plots at various concentrations of norcholine. The electrochemical mechanism of norcholine was physiochemical and which thereafter deviated to chemisorption adsorption at higher norcholine concentrates in consonance with Langmuir and Freundlich isotherms illustrations. The inhibitor molecules selectively precipitate on the reactive areas of the steel exterior in consonance with cathodic inhibition reaction.

Acknowledgement
The author appreciates the support of Covenant University Ota, Ogun State, Nigeria for this project.

References

[1] Uhlig HH, Revie RW 1985 *Corrosion and Corrosion Control* (New York: John Wiley and Sons).
[2] Sastri VS (1998) *Corrosion Inhibitors, Principles and Application* (New York: John Wiley and Sons).

[3] Loto RT, Leramo R and Oyebade B 2018 Synergistic combination effect of salvia officinalis and lavandula officinalis on the corrosion inhibition of low-carbon steel in the presence of SO_4^{2-} and Cl^- containing aqueous environment *J. Fail. Analy. Prev.* **18(6)** 1429–1438.

[4] Loto RT and Olowoyo O 2018 Corrosion inhibition properties of the combined admixture of essential oil extracts on mild steel in the presence of SO_4^{2-} anions *S. Afr. J. Chem. Eng.* **26** 35-41.

[5] Dwivedi D, Lepková K and Becker T 2017 Carbon steel corrosion: A review of key surface properties and characterization methods *RSC Adv.* 7 4580-4610.

[6] Singh WP and Bockris JO 1996 Toxicity Issues of Organic Corrosion Inhibitors: Applications of QSAR Model (Texas: NACE International)

[7] Sykes JM (1990) Silver Jubilee review 25 years of progress in electrochemical methods *Brit. Corr. J.* **25** 175-183.

[8] Elachouri M, Hajji MS, Kertit S, Essasi EM, Salem M and Coudert R 1995 Some surfactants in the series of 2-(alkyldimethylammonio) alkanol bromides as inhibitors of the corrosion of iron in acid chloride solution *Corros. Sci.* **37** 381. https://doi.org/10.1002/chin.199526010

[9] Mernari B, Elattari H, Traisnel M, Bentiss F and Lagrenee M 1998 Inhibiting effects of 3,5-bis(n-pyridyl)- 4-amino-1,2,4-triazoles on the corrosion for mild steel in 1M HCl medium *Corrosion.* **40(2-3)** 391- 399.

[10] Muralidharan S, Phani KLN, Pitchumani S, Ravichandran S and Iyer SVK 1995 Polyamino-benzoquinone polymers: A new class of corrosion inhibitors for mild steel *J. Elect. Soc.* **142** 1478-1483.

[11] Geethamani P, Narmatha M, Dhanalakshm R, Aejitha S and Kasthuri PK 2019 Corrosion inhibition and adsorption properties of mild steel in 1 M hydrochloric acid medium by expired ambroxol drug *J. Bio & Tribo-Corr.* **5** 16.

[12] Loto RT 2017 Study of the synergistic effect of 2-methoxy-4-formylphenol and sodium molybdenum oxide on the corrosion inhibition of 3CR12 ferritic steel in dilute sulphuric acid *Results Phys.* **7** 769-776.

[13] Loto CA and Loto RT 2013 Effect of dextrin and thiourea additives on the zinc electroplated mild steel in acid chloride solution *Int. J. Elect. Sci.* **8(12)** 12434-12450.

[14] Loto RT and Loto CA 2012 Effect of P-phenylediamine on the corrosion of austenitic stainless steel type 304 in hydrochloric acid *Int. J. Elect. Sci.* **7(10)** 9423-9440.

[15] Lai C, Xie B, Zou L, Zheng X, Ma X and Zhu S 2017 Adsorption and corrosion inhibition of mild steel in hydrochloric acid solution by S-allyl-O,O′-dialkyldithiophosphates *Results Phys.* **7** 3434-3443. https://doi.org/10.1016/j.rinp.2017.09.012.

[16] Stango SAX and Vijayalakshmi U 2018 Studies on corrosion inhibitory effect and adsorption behavior of waste materials on mild steel in acidic medium *J. Asian Ceram. Soc.* **6(1)** 20-29. https://doi.org/10.1080/21870764.2018.1439608.

[17] Soltani N, Behpour M, Ghoreishi SM and Naeimi H 2010 Corrosion inhibition of mild steel in hydrochloric acid solution by some double Schiff bases *Corros. Sci.* **52** 1351-1361.

[18] Obot IB and Obi-Egbedi NO 2008 Inhibitory effect and adsorption characteristics of 2,3-diaminonaphthalene at aluminium/hydrochloric acid interface: Experimental and theoretical study *Surf. Rev. Letts.* **15(6)** 903-910.

[19] Obot IB, Obi-Egbedi NO, Umoren SA and Ebenso E 2010 Synergistic and antagonistic effect of anions and ipomoea invulcrata as green corrosion inhibitor for aluminium dissolution in acidic medium *Int. J. Elect. Sci.* **5** 994-1007.

[20] Obot IB, Obi-Egbedi NO and Umoren SA 2009 Adsorption characteristics and corrosion inhibitive properties of clotrimazole for aluminium corrosion in hydrochloric acid *Int. J. Elect. Sci.* **4** 863-877.

[21] Obot IB and Obi-Egbedi NO 2010 Adsorption properties and inhibition of mild steel corrosion in sulphuric acid solution by ketoconazole: Experimental and theoretical investigation *Corros. Sci.* **52** 198-204.

[22] Sahin M, Bilgiç S and Yılmaz H 2002 The inhibition effects of some cyclic nitrogen compounds on the corrosion of the steel in NaCl mediums *Appl. Surf. Sci.* **195** 1-7.

Assessment and Enhancement of the Manufacturing Productivity through Discrete Event Simulation

Naif Alsaadi[1, *]

[1]Department of Industrial Engineering, Faculty of Engineering Rabigh Branch, King Abdulaziz University, Jeddah, Saudi Arabia.

nalsaadi@kau.edu.sa

Abstract. A balanced and resource-efficient assembly line plays a pivotal role in deciding the profit or loss of an industry. In automotive industries, the productivity of the specific materials is highly dependent on several explanatory variables and the processing operations. Moreover, in this 21st century, simulation is considered a powerful and cost-efficient tool for the designing and analysis of manufacturing operations. Therefore, in this research work, a case-study-based assessment of the assembly line in an automotive component vendor company has been proposed through the study of discrete events. Different processing operations are analyzed in terms of the productivity and effective utilization of available resources. The proposed model is set up by taking actual data using time study techniques from each workstation and process along the production line. All the processes, starting from the arrival of raw materials, forging, to the packing of final products are studied and their data is used to set up a model in software. A detailed statistical distribution of the data urged us to simulate the working area in the virtual environment under the different scenarios to define the bottlenecks. The simulation results have elaborated the realistic sense and meaningful contribution towards the performance enhancement of all the processes involved in the production line. Interestingly, Discrete event simulation has helped us to increase productivity as well as reduce the cost by eradicating the less used resources of the assembly line. Based on the results obtained, it is highly recommended to utilize the proposed methodology in the production system for acquiring better-quality and cost-effective products.

1. Introduction

Industries are changing their production systems due to changes in their client's demands. Flexible and faster production systems are necessary to achieve a high variety of products simultaneously in small numbers. There is a need for line balancing to manage the high demand in variety with complex layouts and changing operations on different workstations. To avoid interruptions in the production flow, idle time must be minimized. Common issues in a production line that need to be addressed, include firstly, reducing the operation time given on different workstations. Secondly, minimizing the workload of the operator on each workstation when cycle time and workstation cannot be altered. And thirdly, to improve man and machine resource utilization by reducing the idle time [1].

Changing the layout of the production line can be a costly and tedious task. Likewise, deciding whether a new set of changes will perform better than the current scenario is hard to decide until the new system is established. If the new layout doesn't achieve the required improvements and results, it will turn into wastage of time and capital. So, it is desirable to test the ability of the new production system with a new set of changes before purchasing and installing new equipment [2,3]. The assessment of different staff requirements and few distinct designs can be imitated with ease in the simulation model, which would otherwise have taken years for testing or gathering the data. The ProModel software has been proved to be a powerful tool in assessing the changes for a manufacturing line before incurring actual manufacturing improvements[4]. Omogbai, et al., [5] improved the existing system of manufacturing by using Lean techniques. Lean assessment manufacturing is a tool being used to assess the current performance and after which the predefined system is optimized using discrete event simulation. Changes were suggested and implemented in simulation software to validate the future performance of the proposed design of manufacturing. The objectives included were setup time reduction, quality management, total quality management, reduction of waste, and just-in-time production (JIT). The improvements that were made caused the system to perform better by reducing setup time by 100%, and by increasing preventive maintenance frequency wherein the average time between downtime and uptime was increased to 10,000 minutes. To reduce motion waste, the travel time between workers and machines was reduced to zero. The lead time of the supplier was reduced to zero using all the resources as in the previous system. The changeover time of the machines was reduced by 100%. The system was run at the fastest theoretical time to complete the operation. Different scenarios were created to optimize the performance of the system. The best scenario was then chosen to be implemented, and it showed improvement from the previous system.

El-Khalil, et al., [6] proposed a solution for the inefficiencies that were identified in the assembly line of the car industry using discrete event simulation. One-year data was taken from the vehicle assembly line and used as input for the simulation model. The simulation model was imitated with the real system and the results of the simulation showed all the bottleneck points that were present within the system. Two improved scenarios were tested on the simulation that improved the cost by $1.6 million/annually. The results that were obtained helped the management in making decisions about the size of the buffer and batch size, as well as to determine alternative solutions to solve the bottleneck points for making improvements in the current system. The downtime, maintenance time, labor efficiency, and uptime of robots were not considered in the simulation model.

Liu, et al., [7] used an object-oriented technique of simulation modeling technology to optimize issues such as non-value adding processes, high WIP load, and large cycle resulting from the problems in the system. The achieved objectives were to highlight the bottleneck points and processes causing the large cycle time. The balancing of the line was done using eM-plant simulation software which discovered the operation and process causing limited production and large cycle time. For this purpose, a virtual simulation environment was developed using the simulation software after collecting the data from the assembly line to address the issues facing automobile manufacturing enterprises. The proposed solution for the assembly line developed using the simulation software caused an improvement in the overall efficiency of the assembly line that catered to the interior of the automobile, which resulted in smoother and more adept operations. The improvements made in the production line addressed the issues that were restricting the reduction of the cycle time and causing limitations in the production line.

Dewa, et al., [8] worked on batch modeling lines that were the best fit for moderate demand, although, they were categorized by high WIP inventories, which caused an increase in the product cost and led to a large cycle time of the designed line. Improvements were made to achieve flexibility in the production of the final products when the demand of the clients changes for the different types of the product with different demands. A model of the mixed lines was proposed to reduce the high work in process and improve flexibility to handle the different products at a time with the reduced cycle time. The objectives of the study were achieved for a manual assembly line system for automobiles and the system was optimized by pointing out the bottlenecks and making sure that maximum utilization was achieved for machine and human resources. The cycle time was reduced by reducing the length of the queue before the arrival at the workstation. This also helped in reducing the downtime of the working station. The study was carried out by collecting extensive time study data from the floor of the assembly line. The data obtained from the time study was utilized using a simulation model development of the existing system. The developed model was validated with the current performance of the system. After validation of the simulation model, a detailed analysis of the bottleneck study was conducted, and an optimal solution was suggested for the batch mode of the manual assembly line. The devised methodology that was developed could also be used for a single model of the manual assembly line for the achievement of the objectives.

Marsh, et al., [9] worked on a dynamic model using discrete-event simulation that provided holistic modeling of the manufacturing system, involving simulation of all the processes, operational details cost of each particular resource being utilized. Using discrete event simulation, a complete model of the manufacturing system and its associated costs was developed using simulation software. The production cost was optimized by controlling the lead time, cycle time, and inventory. Many scenarios were developed to optimize the cost of production through the simulation model. Suggestions were developed by analyzing the simulation results of the existing system. Upon running the system after implementing the suggested changes, it was found that the system improved and performed better in the context of the associated cost. The optimized design of the production system performed better by saving the assembly cost.

Savory, et al., [10] worked on the cost estimation of a cellular company that manufactured mobile cells at a small-scale using group technology rather than job technology. A concept activity-based costing (ABC) was introduced in this particular case study using discrete-event simulation. The model was developed on a U-shaped layout wherein the final product was being produced by a part of the family along with its four members. The data for each activity was taken and specific information about the cost of every attribute was considered. The model was enhanced for analysis of cost estimation involving economic considerations. The final model that was developed showed a reduction in invested cost and an increase in profitability.

Holt, et al., [11] focused on an assembly line that was responsible for the installation and assembly of the booms of the truck within the truck bed utility section of the plant. Multiple layout designs were tested to increase the line efficiency and throughput rate of the final products. Material handling was improved, the storage point was reconsidered, and the assembly line was balanced to achieve the predefined objectives. The current system model was simulated using Flexism simulation software for each of the layouts involved to validate the system with the current performance of the system. Scenarios with multiple staffing were developed to determine the optimum layout that would produce maximum productivity. A two-phase layout strategy was developed after analyzing the results obtained using different layouts. The first phase of the layout was implemented for short-term production planning of the industry. The second phase was implemented to meet the demand for future long-term production goals.

Abed [12] discussed process optimization in the food processing industry where there is a problem associated with decision-making to meet the variable demand of the customers while keeping the quality at the highest level. This case study was conducted in a Rusk manufacturing company using Arena simulation software to validate the current performance of the industry. An analysis of the different processes involved in the production process was conducted along the production line. Data was collected for each process for inputting into the simulation model. The current performance of the system was validated, after which what-if scenarios were tested to optimize the existing manufacturing system. The simulation results highlighted the bottleneck points and the under-utilized resources. The suggested changes resolved all the problems and improved the utilization of resources. After the potential changes were implemented, there was a reduction in the cycle time by up to 11.4 percent and an increase in the throughput by 50 percent with an additional investment of capital. The invested capital was monetarily admissible because it was paid back within 35 days by the additional profit that was achieved as a result of the changes made and due to an increase in the productivity of the manufacturing line.

Masood, et al., [13] presented the line balancing of a cylinder block production line to increase the machine resource utilization and reduce the total cycle time of the cylinder block in an automotive production plant. Line balancing of the production line was verified by simulation results. The results indicated that there was a significant increase in machine utilization and throughput rate. Critical problems were identified that caused the highest cycle time and were selected for optimizing the performance of the production system as part of the study. The improvements made in the simulation model resulted in a reduction of the cycle time from 293.9 to 200 seconds, which represented a reduction in cycle time by 32 percent. As a result of the reduction in cycle time, the throughput has increased by 65 percent. At all the stations there was an improvement in utilization by up to 95 percent.

Scholl, et al., [14] solved the problem of balancing an assembly line and redesigned the line through optimization of configuration. The objectives that were to be achieved were workload management at every workstation along the assembly line, reduction in cycle time, re-arranging the layout of the assembly line, and improving the efficiency of the line. Different procedures were tested to optimize the current assembly line. The best method was implemented, and changes were made to achieve the objectives.

Yu, et al., [15] used simulation modeling using computer technology to predict the performance of the manufacturing assembly line even before the production process itself is initiated. The system was planned to keep in view ergonomic design, and human and machine utilization. The prime focus was the productivity improvement of the assembly line. The shortcomings in the system were identified after the validation of the current performance using simulation software. Data was collected from each workstation of the assembly floor regarding the devices which assisted in lifting, for being put into the simulation software. The productivity of the simulation model was validated with the current manufacturing system. Human resource and machine utilization results were obtained from the output of the pro-model software. The problems of the assembly line were optimized, and an improved model was developed.

Baesler, et al., [16] presented the model of a sawmill using discrete event simulation. Data was taken from all the workstations for building the simulation model. The simulation results of the current system validated the productivity results of the proposed model. Bottleneck analysis of the woodwork was done to identify the processes that negatively impacted the productivity of the production line. Different alternative solutions were tested to yield better production with improved process utilization. The results obtained by every design combination were compared and the best set of improvements was proposed for implementation to improve the productivity of the production line. The implementation of the changes to the improved model increased the throughput by up to 25 percent.

Samuel, et al., [17] considered traditional parameters of productivity metrics such as throughput and rate of resource utilization to be insufficient for evaluating the performance of the manufacturing system to bring about improvements in the system. A systematic methodology was developed for the measurement of productivity and factory-level analysis of the system. Equations for the productivity parameters of Overall Throughput Effectiveness (OTE) and Overall Equipment Effectiveness (OEE) were developed for the productivity measurement of the system and the quantitative measurement of its equipment. These parameters were calculated using simulation software and the productivity of the system was validated. After validation, improvements were made in the system. A simulation model was developed through data collection of all the processes at the factory level. The results obtained by the pro-model software showed that the improved system behaved better as compared to the previous system.

The fundamental aim of this research work is the implementation of the discrete-event methodology on the realistic data of the automotive industry for enhancing throughput efficiency. A statistical distribution summary is included to acquire concise information on the dominant manufacturing operations for the production of automotive components. It is noticed that conventional assembly lines have less utilization of the available resources and consequently they increase the idle in a system. Therefore, a time-series technique is utilized to gather the relevant data from the workstations of the automotive industry. A conventional statistical distribution has helped us to normalize the data and therefore, a process simulator is used to simulate the working environment under realistic predefined conditions. The duration of one month is selected to analyze the productivity of the assembly line and it is observed that simulation results are highly correlating the realistic production with a percentage error of less than 12%. However, results revealed that different processing operations decrease the efficiency of the production system. It is due to less utilization of the available resources, therefore, two different scenarios on the predefined working assembly line are implemented to increase the productivity of the automotive industry. Based on the outcomes of the research work, it is highlighted that the proposed simulation technique can also be utilized to balance the assembly line of the automotive manufacturing industry.

2. Simulation and Analysis

In this section, a concise description of the integration of Discrete event simulation with the automotive industry is presented to highlight the importance of numerical assessment in the manufacturing industry. Additionally, a detailed explanation about the implementation of the simulation model in the real-time working environment is included to signify the outcomes of the proposed methodology.

2.1. Integration of Discrete Event Simulation with Automotive Industrial Materials

In the present era, several categories of composite materials are being utilized in the automotive industry to improve the performance of several machines particularly motor vehicles. Interestingly, the properties of these key engineering materials are highly dependent on time; therefore, processing operations and assembly are carefully designed to meet the benchmark. Although, several techniques are also available to continuously improve the product development and efficiency of the assembly line. However, researchers have claimed that those techniques are not cost-efficient in terms of realistic implementation on the predefined data. Additionally, in the recent past, it is highlighted that piezoelectric materials would be the best option for smart automobile products because of their inherent nature of the direct and indirect exhibition of electrical charge and deformation, respectively. However, the integration of such materials with the sub-components of the product is solely dependent on the assembly line and production layout. Over the past few years, researchers have demonstrated the influence of lead-time and processing operations on the mechanical quality factor of the respective engineering materials. However, decision-making variables are rarely identified because of critical problems that occurred during manufacturing operations. Therefore, researchers have been recommending to integrate the discrete event methodology with the automotive industrial materials to eradicate the possible errors and enhance the product efficiency Additionally, the realistic simulation model of the predefined system would help the researchers to utilize the available resources efficiently. In this research work, discrete event simulation has been implemented on the assembly line comprising of time-variant processing operations for the development of a typical automobile product. It is highly recommended to utilize the proposed methodology for maintaining product quality in a cost-efficient way.

2.2. Implementation of the simulation model on the Working Assembly Line

The behavior of a system can be easily imitated using simulation software to help researchers in decision-making and authorities in bringing about improvements in their production system. This process is accomplished by primarily conducting a time study of that system. After fully duplicating the system, the software-based results can help in predicting the possible risks of failure of a decision related to the system. A lot of money and time wastage can be prevented if the system is properly studied. In this study, the bottleneck points in the production line are to be identified and possible solutions are suggested to assist decision-making at the managerial level. After the time study of each workstation, the data has been put into the statistical analysis software to find out the most suitable combination of best-fit distribution that can be put into the simulation software.

The structural outline of the simulation study has been illustrated in Fig 1. The objective that has been set for the system under discussion is to address the problems in the production system. For the study, the production system was imitated in the simulation interface and first validated for current performance and output. The output of the system after validation indicates the problems in the assembly line. All the attributes of the production system have been simulated in the software including the warmup time of the system, the number of workstations, setup time, the arrival time of entities involved, resources attached to the workstations, and their availability time. A calendar was programmed with the shift timings of the industry and holiday schedule. The simulation model was developed in such a way that it behaved like a real working system. Once all the time study data was submitted to their respective places, the model was then ready for simulation. The simulation was run after setting the simulation time of one month. The output viewer of the system showed the results, which were presented in the form of graphical bar charts and tables listing the attributes. Once the production system was validated, improvements were tested for the solution of the problems present in the current scenario of the production system. After the improvements were tested, the results were compared, and the best result scenario was chosen for decision implementation.

IOP Conf. Series: Materials Science and Engineering 1222 (2022) 012011 doi:10.1088/1757-899X/1222/1/012011

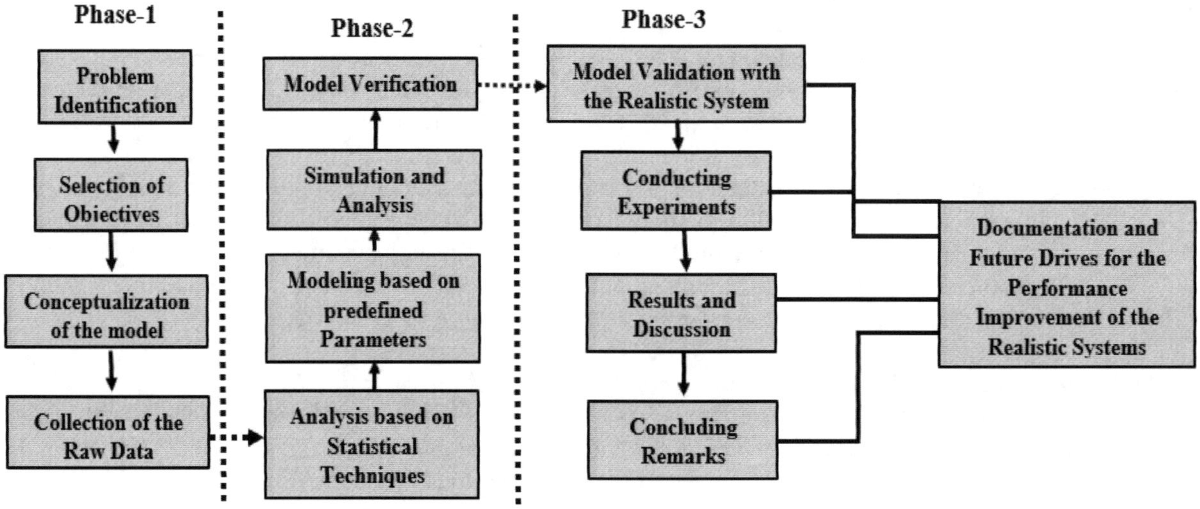

Figure 1. Proposed Research Methodology.

3. Developing the Current Scenario and the Assumptions

An outline has been drawn before simulating using the software to save plenty of time while developing an imitation of the production system. In the outline, all the resources, workstations, buffers, inventories, and locations have been designed according to the flow of the product. All the entities are situated at their proper place and locations according to the real industrial layout. All the operation times and travel times used in the simulation are relevant and from reliable sources. Operations used in the outline are shown in Table 1.

This simulation setup uses assumptions similar to every simulation software due to the availability of insufficient data. The assumptions are made for short periods until accurate readings are available. However, it is interesting to note that the output of the system is not affected if the assumptions are minor. Another approach for dealing with assumptions is to create virtual situations of worst and best scenarios to overview the impact of the assumptions on the system. In this study, the following are the assumptions that are to be considered in the modeling of the production system.

- Activities designed are dynamic.
- The attendance of the working resources has not been considered in the calendar.
- The time required to set up the machine for each process is constant.
- Only one worker can work on each workstation.

Table 1. Operations Used in Simulation Model.

1	Arrival	Arrival type is scheduled at 20,000 pieces per order. A sample for inspection is taken from the lot by the inspector and checked for the quality of forging upon arrival.

2	Inspection	The Inspector takes a sample of 20 pieces for visual inspection, dimensional inspection, and fixture (profile) inspection. After it, he compares the sample with the standard dimensions to decide whether the lot is acceptable or not.
3	Grinding	The pieces are dried and then ground with sandpaper from all the sides of the peddle.
4	Turning	The turning process is done to the workpiece and the tip to make the surface round using the lathe machine. The product is moved to the next operation through the same bucket as a batch of 100.
5	Drilling 1	A 12 mm through-hole is bored by the drill machine operator by controlling the machine manually. The rejects from the drilling process are dumped in the drum placed nearby. The drilled pieces are moved manually to the next operation in a bucket as a batch of 100.
6	Milling	A cut on the end of the tip is machined with a manual vertical milling machine. The operator accumulates the parts and then moves to the next machine since the chamfer in the drilled hole is operated by the same operator.
7	Facing	This process of removing metal from the end of a workpiece is done to produce a flat surface.
8	Chamfering 1	This operation is done to make a bevel on the workpiece.
9	Burnishing	A smooth hard tool is rubbed with significant pressure on the metal surface.
10	De-burring	This is done to remove the burrs over the surface and make the workpiece smoother.
11	Chamfering 2	Again, this operation is done to make a bevel on the workpiece.
12	Fastening Cut	The cut is made using a milling machine for fastening purpose.

The layout of the production system is illustrated in Fig. 2.

IOP Conf. Series: Materials Science and Engineering 1222 (2022) 012011 doi:10.1088/1757-899X/1222/1/012011

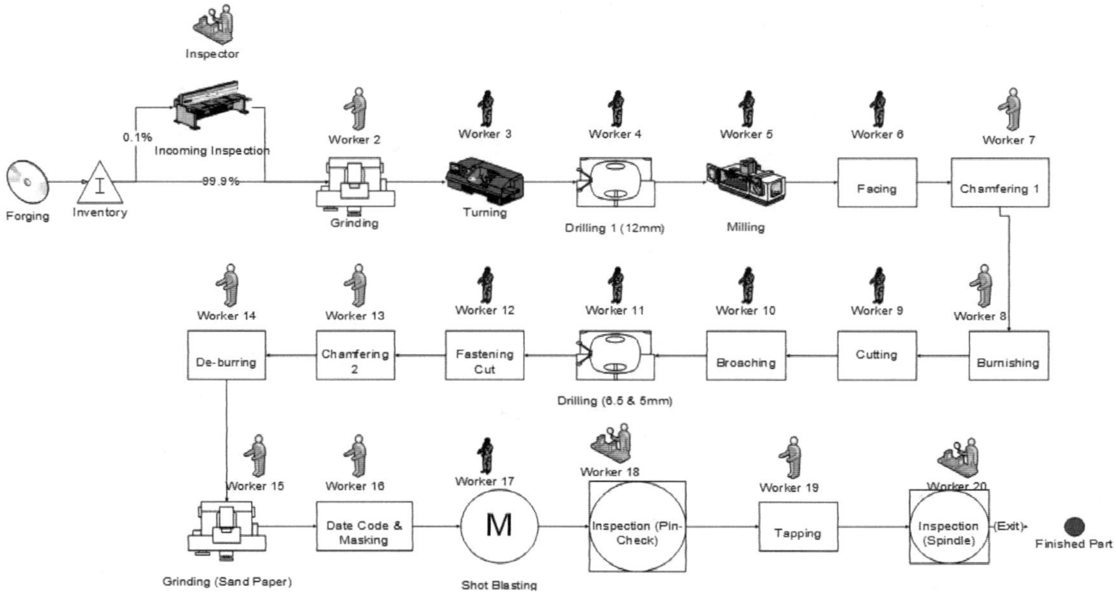

Figure 2. Production Layout.

The schematic illustrates the layout of the industry and the processes involved in the production of automotive components. The 0.1 % of incoming forging is inspected by the inspector to testify the quality of the forging lot. In case of a successful inspection, the forging lot is accepted, and it flows towards the next workstation until it reaches the final product. The main focus of the study is to find the bottleneck point in the production line. The bottleneck is the point of congestion in the assembly line which limits or slows down the production system. It limits the capacity of the whole manufacturing system. Bottlenecks happen when the worker is slow down due to overload or as a result of the parts or material arriving too quickly to his/her workstation. As a consequence of the bottleneck point, the overall cost of production increases. A bottleneck refers to the characteristics of a bottle, whereby the neck of the bottle limits the flow of the liquid due to it being the narrowest point of the bottle. This causes congestion and slows down the liquid flow. Similarly, the bottleneck point slows down the production of forthcoming stations consequently, increasing the time and resources required for producing the final product. It is a critical issue for the manufacturing industries because the bottleneck point is caused by a flaw in the production process due to which it must be corrected. When the demand increases, this issue becomes more prominent and it limits the capacity of the production system.

To simulate the current system of the production line, the data is collected from each process and imitated using simulation software. After all the processes are simulated the warmup time is input to get more precise results. The calendar is included for every resource and machine that contains all the shift timings, working days, and holidays.

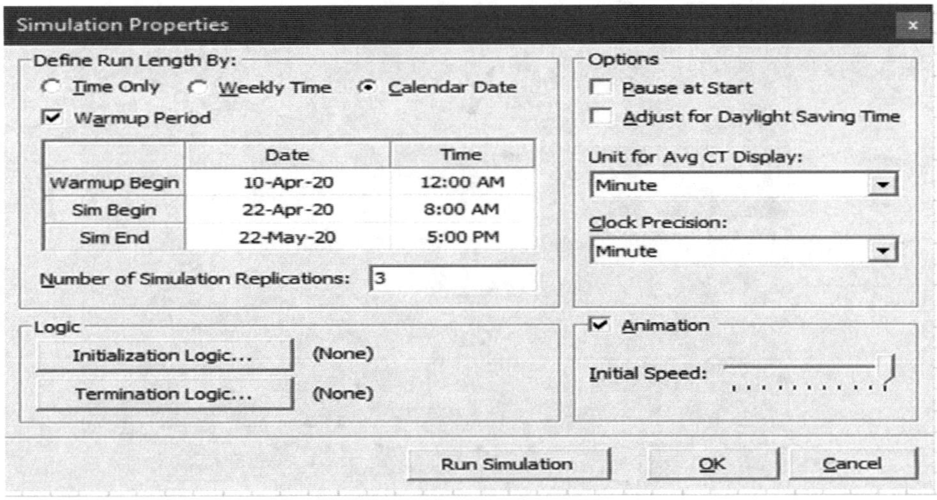

Figure 3. Simulation Properties.

The warmup period is selected in such a way that all the processes are activated, and the workpiece reaches all the workstations. The number of replication or repetition is set to be 3, to get more precise results. The simulation is run for a month to get the monthly production as depicted in Fig. 3. Firstly, the current system is validated, and the bottleneck points are indicated. Secondly, the different scenarios are proposed to overcome the problem. All the required data is collected, and it is analyzed to develop statistical distributions that can be equipped in the activity time. The layout is designed such that it fills the required data boxes, whereas the complex activities that require logic are developed using a set of logic in the logic builders. A calendar with shift timings is added for every activity and arrival.

The output viewer comprehends the graphical analysis of all the attributes of an entity. The resource utilization and machine utilization graph are available in the form of horizontal bars and vertical bars. Time plots, histograms, and pie charts are generated automatically in the output viewer of the Process Simulator. The output viewer reveals that the total exit is 9,166 and the resource utilization of the respective workers is indicated in graphical form. The resource utilization for the current system is shown in Fig. 4.

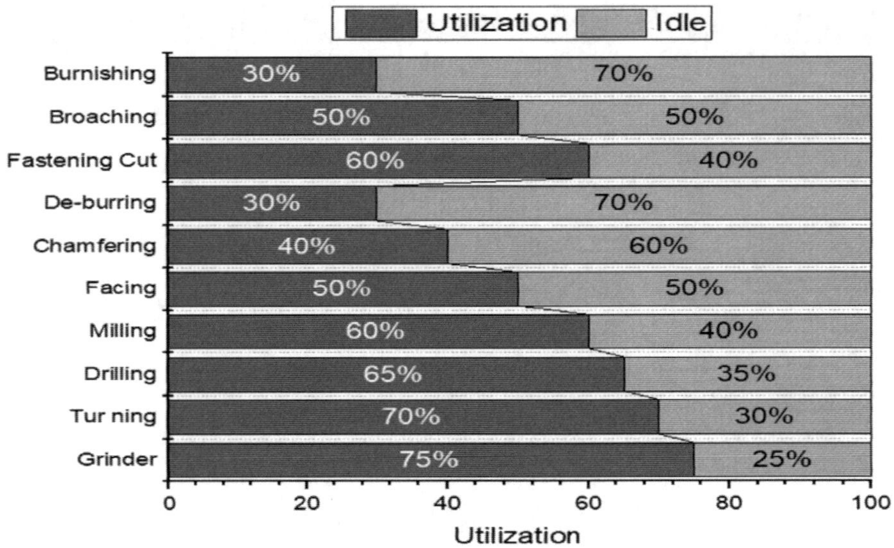

Figure 4. Resource Utilization.

4. Results and Discussion

Analysis of the numerical model shows that most of the resources are being wasted in the assembly line and consequently, the productivity of the automotive industry is less efficient and costly. It is observed that before the implementation of the discrete-event methodology on the plant, idle time and block time are the two dominant variables that adversely affect the system performance as shown in Fig 5. It is because of the insufficient usage of the available resources. Due to these reasons, certain time delays usually appear in the production system which in turn affect the quality as well as the lean time of the production system.

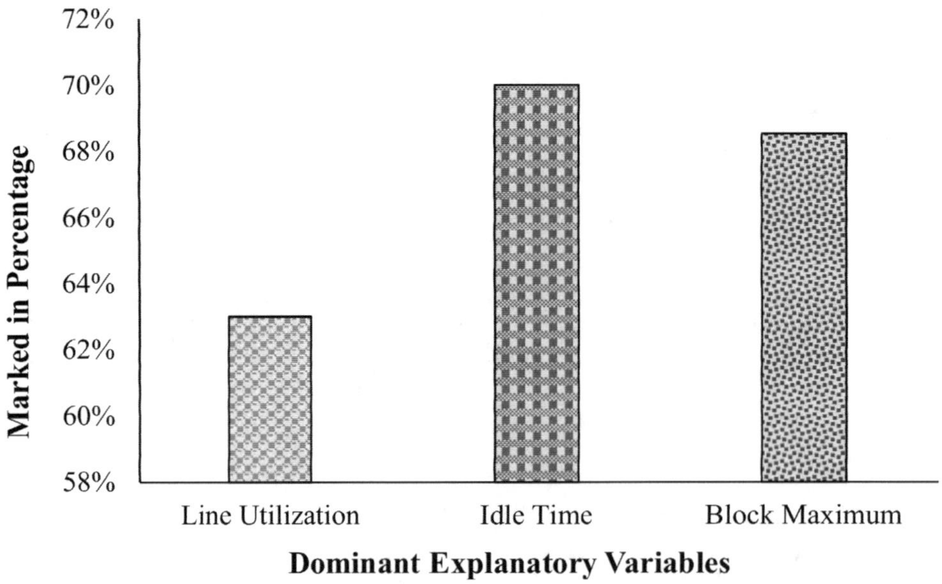

Figure 5. System Performance before implementation of the proposed methodology.

Therefore, the proposed methodology is implemented and the simulation results indicate that the productivity of the assembly line is 9,166, which is nearly equal to the average production of the system (~9,300). Hence there is a negligible difference between the average production value and the production value obtained from the simulation. Additionally, two different scenarios are developed using scenario managers to increase the production of the line as indicated in Table 2.

Table 2. Developed Scenarios.

S. No.	Description	Change
1	Removal of the resources with low utilization and shifting his job to other resources to boost up the utilization. Workers 16 & 19 at Tapping and Date Coding are removed, and their task is assigned to another worker to improve the resource utilization.	Capacity, Production and Utilization Improved, Cost saved
2	Adding a machine in parallel at the bottleneck point of drilling operation as indicated by simulation results, as shown in Fig 6.	1→2

The first scenario is developed without causing the expense of any capital investment. In this scenario, the resources that were being less utilized are removed from the production line and shifted somewhere else while their task is assigned to other workers which cause a significant increase in the utilization of their skills. Additionally, in the second scenario, capital investment is implied because of the integration of the secondary machine in the production layout. The reason behind adding a machine resource is because simulation results proved that, it is a bottleneck process that limits the throughput of the assembly line. The effect of the different scenarios on the drilling operation is shown in Fig. 6. Interestingly, by implying the two time-variant scenarios, efficient performance in the production system is noticed. Additionally, it is noticed that the proposed methodology enables the utilization of the available sources in an efficient way which indirectly increases the quality of a typical product. Moreover, scenario-1 shows more blockage of the products which therefore slows down the forthcoming operations. Therefore, scenario-2 is implemented to eradicate the blockage for the better performance of the production system. It can be anticipated that adding a machine parallel to the existing machine can decrease the blockage and improve the overall system.

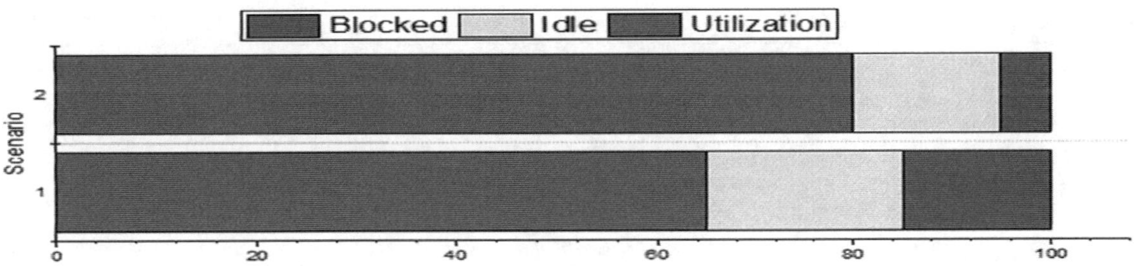

Figure 6. Drilling Operation.

IOP Conf. Series: Materials Science and Engineering 1222 (2022) 012011 doi:10.1088/1757-899X/1222/1/012011

The simulation results disclosed that after potential changes were made in the assembly line, the production of the exited quantities is adequate for the industry to meet the demand of the customers without any overtime expenditure. After applying the suggested alterations and running the improved *scenario 1* through simulation, the product produced at the exit of the assembly line is 9,766 units as shown in Table 3, and resource utilization increases as shown in Fig. 7. This indicates that the demand of the customers can be meet by merely implementing the proposed changes without making any capital investment.

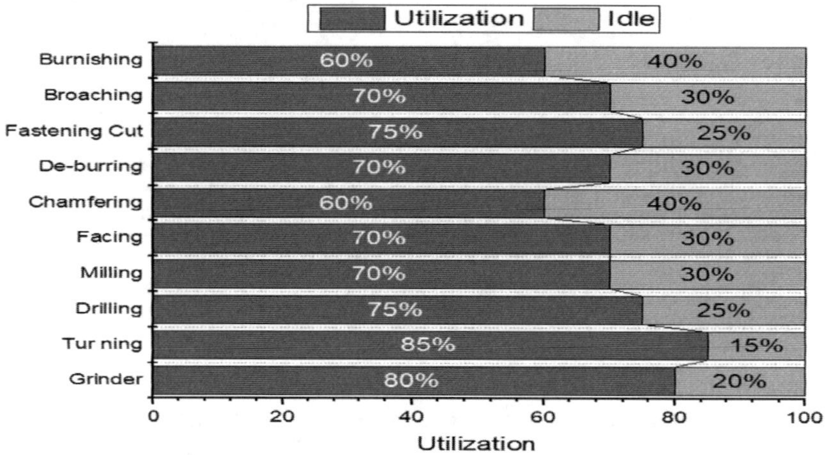

Figure 7. Resource Utilization after Improvements.

Additionally, outcomes of the scenario-2 justified that the productivity of the system increases and efficiently matches the user requirement. It is also noticed that simulation highlights the region in which bottleneck lies and by investigating the possible reasons, those uncertain errors can be minimized to an acceptable level. Moreover, implementing the proposed scenarios not only enhances the throughput but also plays a pivotal role in diminishing the idle time of the processing operations.

Table 3. Results After Improvements

S.No.	*Scenarios*	*Effect on Productivity*	*%Age Increase*	*Effect on Utilization*
1	Removing Human Resource	9,766	05%	45%
2	Adding a Machine	18,064	49%	77%

Adding a machine could also be monetarily admissible since it brings about an improvement in productivity by a big margin which helps the company in meeting the customer's demand and producing more than the demand to achieve higher profitability. Some other improvements made had no direct effect on the throughput of the assembly production line, but they caused the load to become balanced for other operations and workers by increasing their utilization.

Fig.8 elaborates the comparison of the productivity in terms of pieces per month for three different situations. It is noteworthy to analyze that the productivity of the predefined working environment has increased after the implementation of the proposed scenarios. Implementation of the simulation model with the secondary machine (Scenario-2) produces more dominant output in terms of productivity as compared to the model without capital cost (Scenario-1).

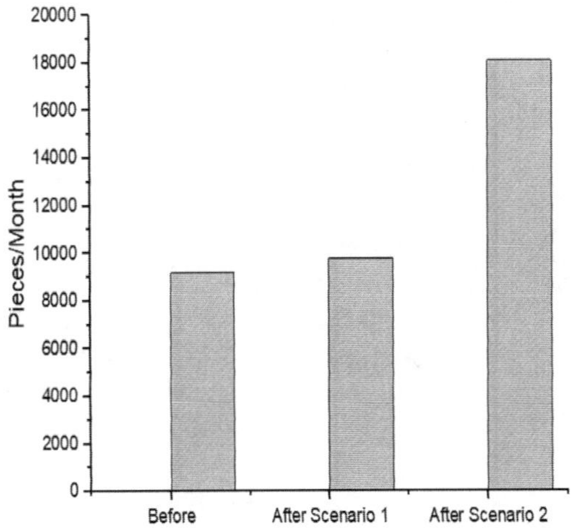

Figure 8. Production Comparison.

5. Conclusion

Manufacturers are continuously trying to improve the dynamics of the production system by implying several useful methodologies. However, one of the major disadvantages of those methodologies is the realistic implementation in a particular environment. Therefore, in this research work, a detailed methodology is presented to highlight the significance of the simulation model in the automotive industry. A sequence of time-series data is utilized to visualize the statistical distribution of the assembly data. After it, a virtual environment is created on the process simulator to analyze the working environment under the different realistic explanatory variables. A concept of discrete event simulation is implemented on the predefined system with two different scenarios of the operating machines. Results of the case-study-based assessment highlighted those discrete events are useful in evaluating the performance of a typical assembly line. Additionally, it is also observed that before the implementation of the proposed methodology, resources are not utilized conveniently which in turn increases the idle time in the processing operations. However, after the implementation of the simulation model in the working environment, the productivity of the system increases exponentially. Based on the outcomes of the proposed methodology, it is highly recommended to imply the discrete event simulation in the automotive industries for the efficient development of better-quality products.

References

[1] Zhao, C., Kang, N., Li, J., & Horst, J. A. (2017). Production control to reduce starvation in a partially flexible production-inventory system. IEEE Transactions on Automatic Control, 63(2), 477-491.

[2] Gopalakrishnan, M., Skoogh, A., & Laroque, C. (2013, December). Simulation-based planning of maintenance activities in the automotive industry. In 2013 Winter Simulations Conference (WSC) (pp. 2610-2621). IEEE.

[3] Williams, E. J., & Ülgen, O. M. (2012). Simulation applications in the automotive industry. In Use Cases of Discrete Event Simulation (pp. 45-58). Springer, Berlin, Heidelberg.
[4] Nallusamy, S., & Ahamed, A. (2017). Implementation of lean tools in an automotive industry for productivity enhancement-A case study. In International journal of engineering research in Africa (Vol. 29, pp. 175-185). Trans Tech Publications Ltd.
[5] Omogbai, O., & Salonitis, K. (2016). Manufacturing System Lean Improvement Design Using Discrete Event Simulation. *Procedia CIRP, 57*, 195-200. doi:10.1016/j.procir.2016.11.034.
[6] El-Khalil, R. (2015, 02). Simulation analysis for managing and improving productivity. *Journal of Manufacturing Technology Management, 26*(1), 36-56. doi:10.1108/jmtm-03- 2013-0024.
[7] Liu, S., Jiang, S., & Hou, K. (2012). An Analysis on the Automobile Interior Assembly Line Balancing Based on eM-plant. *Proceedings of the 2nd International Conference on Computer Application and System Modeling.* doi:10.2991/iccasm.2012.1.
[8] Dewa, M., & Chidzuu, L. (2013, 09). Managing Bottlenecks In Manual Automobile Assembly Systems Using Discrete Event Simulation. *The South African Journal of Industrial Engineering, 24*(2), 155. doi:10.7166/24-2-567.
[9] Marsh, R., Jonik, M., Lanham, J., Cheung, W. M., Newnes, L. B., & Mileham, A. R. (2010, 03). Modelling an assembly process using a close coupled generative cost model and a discrete event simulation. *International Journal of Computer Integrated Manufacturing, 23*(3), 257-269. doi:10.1080/09511920903529248.
[10] Savory, P. (2010, 06). Estimation of cellular manufacturing cost components using simulation and activity-based costing. *Journal of Industrial Engineering and Management, 3*(1). doi:10.3926/jiem.2010.v3n1.p68-86.
[11] Holt, R., Simmons, L., Walden, C., Dennis, G., & Hill, T. (2010). "Layout Analysis using Discrete Event Simulation: A Case Study". In *Proceedings of the 2010 Industrial Engineering Research Conference.*
[12] Abed, S. Y. (2008). "Improving Productivity in Food Processing Industries Using Simulation - A Case Study" *12th WSEAS International Conference on SYSTEMS, Heraklion, Greece, July 22-24, 2008.* doi:10.31274/rtd-180813-6046.
[13] Masood, S. (2006, 01). Line balancing and simulation of an automated production transfer line. *Assembly Automation, 26*(1), 69-74. doi:10.1108/01445150610645684.
[14] Scholl, A., & Becker, C. (2006). State-of-the-art exact and heuristic solution procedures for simple assembly line balancing. *European Journal of Operational Research, 168*(3), 666-693. doi:10.1016/j.ejor.2004.07.022.
[15] Yu, Q., Duffy, V., Mcginley, J., & Rowland, Z. (2006). Productivity Simulation with Promodel for an Automotive Assembly Workstation Involving a Lift Assist Device. *Proceedings of the 2006 Winter Simulation Conference.* doi:10.1109/wsc.2006.322977.
[16] Baesler, F., Araya, E., Ramis, F., & Sepulveda, J. (2004). The Use of Simulation and Design of Experiments for Productivity Improvement in the Sawmill Industry. *Proceedings of the 2004 Winter Simulation Conference, 2004.* doi:10.1109/wsc.2004.1371452.

[17] Samuel H. Huang , John P. Dismukes , J. Shi , QI Su , Mousalam A. Razzak , Rohit Bodhale & D. Eugene Robinson (2003) "Manufacturing productivity improvement using effectiveness metrics and simulation analysis", *International Journal of Production Research,* 41:3, 513-527, DOI: 10.1080/0020754021000042391.

Influence of design parameters on the fresh and durable properties of self-compacting concrete with recycled aggregate

R Martínez-García [1,a*], P Jagades[h2], F J Fraile-Fernández[1] and J de Prado-Gil[1]

[1]Department of Mining Technology, Topography and Structures, Campus of Vegazana s/n, University of León, 24071 León, Spain
[2]Department of Civil Engineering, Coimbatore Institute of Technology, Coimbatore 641014, Tamil Nadu, India;

[a]rmartg@unileon.es, [b]jagadesh.p@cit.edu.in, [c]fjfraf@unileon.es, [d] jdeprg00@estudiantes.unileon.es

Abstract. The objective of this paper is to analyse the influence of design parameters on the fresh and mechanical properties in the manufacture of self-compacting concrete with recycled fine and coarse aggregates. Design parameters such as water-cement ratio, water/paste ratio, total aggregate to cement ratio, fine aggregate to coarse aggregate ratio, water to solid ratio in percentage, superplasticiser content, RCA replacement percentage and RFA replacement percentage are analysed. The design parameters are found to affect the evolution of the fresh and mechanical properties of SCC with recycled aggregates.

Keywords: self-compacting concrete; recycled aggregates; design parameters; fresh concrete properties; mechanical properties; mix design.

1. Introduction

One of the most widely used products in the world today are aggregates, specifically sand and gravel, which are used in construction. After water, concrete is the most consumed product on the planet.

Therefore, there is evidence of an increase in the generation of concrete waste related to the construction process and demolition activities [1],[2]. The development of civil works generates a large amount of waste that can be reused that could be used as landfill for the development of new civil works or disposed of in municipal landfills competing for space with municipal solid waste when managed in the same facilities [3], or improperly abandoned generating an environmental impact by the loss of landscape quality in different localities, and environmental pollution [4], [5].

In terms of figures, it should be noted that (1) for the year 2010, worldwide 36% of waste corresponded to construction and demolition [6]; (2) the global average generation of special construction and demolition waste corresponds to 1.68 kg/inhabitant/day in 2018. Based on this reality, researchers and specialists from universities and research and quality control centres around the world have focused on providing management solutions. In addition, several countries have adopted legislation to promote and incentivise the reuse of waste; this has been reflected in the statistics. The amount of waste recovered in the EU has been increasing; it has risen from 45.4% in 2014 to 53.2% in 2018. 37.8% of the waste generated in the EU in 2018 was recycled, 9.9% was used as landfill and 5.6%

was used for energy recovery. The remaining 46.8% was landfilled (38.8%), incinerated without energy recovery (1.0%) or otherwise disposed of (7%). In this context, the long-term European Strategy is to turn the world into a recycling society, avoiding waste and using it as a resource wherever possible. The aim is to achieve much higher levels of recycling and to minimise the extraction of additional natural resources.

Currently, there are numerous studies that support the use of construction and demolition waste in the production of conventional concrete [7–11]. However, despite the increasing trend in the reuse of demolition waste; it has been shown that the use of recycled aggregates (RA) can limit or decrease the properties of concrete (strength, durability and serviceability) due to old binding mortar, irregular shape and increased porosity. In other words, there is a direct dependence between concrete properties and the degree of substitution used with RA [12], [13]. Several tests have concluded that the incorporation of recycled products (coarse aggregates) should not exceed 20% to preserve the properties of recycled concrete, which is a worldwide normative recommendation. In the case of fine material, although it is not regulated or permitted, scientific studies have shown that there is no alteration when it does not exceed 10% substitution, and it can improve with the granulometric gradation in the addition of fine material, generating greater cohesiveness [14], [15], [16].

Given the high energy consumption of the concrete industry due to the high demand for natural resources and the environmental risks involved, RA have been used as partial or full substitutes to minimise the environmental impact, taking an important step towards establishing a sustainable society and boosting the circular economy [12], [17] , [18]. RA is a type of composite material whose quality depends directly on the origin of the waste, the place and reason for demolition and the original characteristics of the concrete [19]. Therefore, it should not be lost sight of the fact that any change in the aggregate, however small it may seem, will have a significant effect not only on the properties of the concrete but also on the cost of production, transport and marketing.

Regarding the relationship of the mechanical properties of concrete (compressive and flexural strength) obtained with RA, some studies have been carried out, using different substitution percentages, finding for example that the use of varying amounts of aggregates between 10%-40% by weight (waste from a 25 year old building) generates a decrease in properties with the most favourable value being 30% by weight substitution [20]. It is also indicated that with up to 40% by weight substitution of recycled coarse aggregate (RCA) and recycled fine aggregate (RFA), a self-compacting structural concrete is obtained without significant differences with conventional concrete [21]. Compressive strength, tensile strength, flexural strength and elasticity have developed well even with 100% substitution, as a consequence of the improved microstructure. On the other hand, hardening properties, stiffness and flexural toughness may decrease with the addition of aggregates [22]. In addition, self-compacting concrete (SCC) usually has a higher water absorption than normal concrete due to the lower specific gravity of the old mortar that binds to the original concrete. Also, the density of SCC with aggregates is lower than that of conventional concrete due to its high porosity [20]. Due to the need to optimise the management of solid waste associated with construction activities and the demand for materials that, in addition to being more sustainable, have adequate mechanical properties that guarantee high durability, research has been carried out on the mechanical properties (e.g. compressive and tensile strength) of SCC obtained conventionally or with the use of RA, using different raw materials, design parameters and verification tests; Therefore, it is necessary to carry out an updated general review to compile the advances achieved so far in this area of knowledge that will serve as a substantial contribution to the establishment of criteria for the implementation in civil works, in the establishment or adaptation of existing regulations.

The main objective of this review is to understand the role of design parameters on the fresh and mechanical properties of SCC with recycled aggregates (RCA and RFA). For this purpose, more than 250 SCC mixes with RA from the available literature have been analysed and grouped into 6 families according to their compressive strength of control mixes.

2. Material properties of SCC ingredients

The origin of SCC can be diverse, but there is a definite role of ingredients properties on the SCC fresh concrete properties. Hence, it is necessary to understand the properties of ingredients used in SCC.The average value of ingredients used for SCC other than RA by various researchers is tabulated in Table 1. From Tables 1 and 2, it is observed that the physical properties of natural aggregate are higher than that of the RA due to old mortar adhering to it. The specific gravity of RA is lower than that of natural aggregate is observed by most researchers, this leads to a density of RCA mixture as lower than that of the natural aggregate mixture. Higher water absorption of RA is observed from Table 2 when compared to Table 1 and it is reported by most authors.

Table 1. Average value for the properties of ingredients other than recycled aggregated used in the literature.

Ingredients	Specific Gravity	Fineness modulus	Blaine's surface area	Water absorption	Bulk density	References
Cement	3.108		3495.5			Tuyan et al., 2014 [23], Pereira-de-Oliveria et al., 2014 [24], Señas et al., 2016 [25], Duan et al., 2020 [17], Tang et al., 2016 [26], Djelloul et al., 2018 [27], Sasanipour et al., 2019 [28], Aslani et al., 2018 [29], Ouldkhaoua et al., 2020 [30], Nieto et al., 2019 [31], Gupta et al., 2020 [32], Singh et al., 2019 [2], Abed et al., 2020 [33], Xavier et al., 2020 [34], Dapena et al., 2011 [35], Rajhans et al., 2011 [36]
Fine aggregate	2.651	2.76		1.35	1632.5	
Coarse aggregate	2.646	6.05		1.24	1440.25	

Table 2. Recycled aggregate properties used in the literature.

References	Source	Types	AS[a] (mm)	FM[b] (mm)	CF[c] (%)	SG[d]	W[e] (%)	BD[f] (kg/m^3)
Tuyan et al., 2014 [23]	Waste laboratory made concrete	RCA	-	-	-	2.48	4.80	1410
Pererira-de-Olivera et al., 2014 [24]	Construction and demolition waste	RCA	9.5	5.78	-	-	4.10	1509
		RCA	19	6.92	-	-	4.05	1485
Señas et al., 2016 [25]	Construction work	RCA	12.5	3.77	-	-	-	-
Duan et al., 2020 [17]	Waste infrastructure component	RCA	12	2.59	-	-	6.53	1220
Tang et al., 2016 [26]	Construction and demolition waste facility	RCA	10	-	-	-	7.75	1450
Djelloul., 2018 [27]	Waste from laboratory concrete	RFA	-	3.8	-	2.27	8.87	1258
		RCA	9.5	-	-	2.39	7.39	1172
		RCA	12.5	-	-	2.4	3.21	1154
Sasanipour et al., 2019 [28]	Residential and sanitary buildings	RCA	-	-	-	2.36	5.4	-
		RFA	-	3.67		2.23	14.8	-
Abed et al., 2020 [33]	Demolition waste	RCA	-	-	-	-	5.60	1612
Nieto et al., 2019 [31]	Concrete, bricks, asphalt, glass, others, aggregates	RCA	16	-	-	-	6.61	1584
Gupta et al., 2020 [32]	Demolished building	RCA	-	-	-	2.54	2.33	1220
Ouldkhaoua et al., 2020 [30]	Cathode ray tube funnel glass	RFA	-	2.16	-	2.75	-	-

[a] Aggregate size (mm)
[b] Fineness modulus
[c] Content of fineness (%)
[d] Specific gravity
[e] Water absorption (%)
[f] Bulk Density (kg/m^3)

Table 3. Average of reference mix proportions (control mix) used by several authors from 2010-2020.

Ingredients (Kg/m^3)	Average values	Standard deviation	References
Cement	386.31	83.64	Tuyan et al., 2014 [23], Pereira-de-Oliveria et al. [24], 2014, Señas et al., 2016 [25], Duan et al., 2020 [17], Tang et al., 2016 [26], Djelloul et al., 2018 [27], Sasanipour et al., 2019 [28], Aslani et al., 2018 [29], Ouldkhaoua et al., 2020 [30], Nieto et al., 2019 [31], Gupta et al., 2020 [32], Singh et al., 2019 [2], Abed et al., 2020 [33], Xavier et al., 2020 [34], Dapena et al., 2011 [35], Rajhans et al., 2011 [36]
SCM	191.4	80.84	
Filler	166.5	182.37	
Fine aggregate	823.95	214.93	
Coarse Aggregate	749.26	113.77	
Water	194.07	23.52	
SP	3.62	1.67	
Unit weight	2335.61	46.12	

3. Results and discussion

3.1 Influence of design parameters on fresh concrete properties

For each fresh SCC properties and every author, the results/findings obtained are presented in the form of a table/graph and the new tendencies are identified and studied based on the design parameters obtained from the mix design. A comparison for similar kinds of experiments is made from various authors to understand the RCA trends on fresh properties. The researchers concluded that several parameters are used to evaluate the slump flow which is defined as free fluidity and flow in the absence of barriers. Parameters considered are the time required by SCC takes to form a 500mm circle, called flow time (T500), and the slump flow (SF) diameter. A common observation made by most researchers is that increase in replacement of coarse aggregate by RA results in T500 increases with SF decreases (Figure 1). Safiuddin et al. [37] reported that the increase in incorporation ratio results in increased segregation resistance up-to lower replacement levels of NCA. For higher replacement levels of NCA, the segregation resistance decreases because the increase in RCA content leads to an increase in the fineness content of the mixture. Grdic et al. [38] found that the increase in RCA content leads to an increase in the W/C ratio, result in a decrease in segregation resistance. An increase in replacement percentage of natural aggregate by the coarse aggregate result in an increase in flow time is observed in Figure 2.

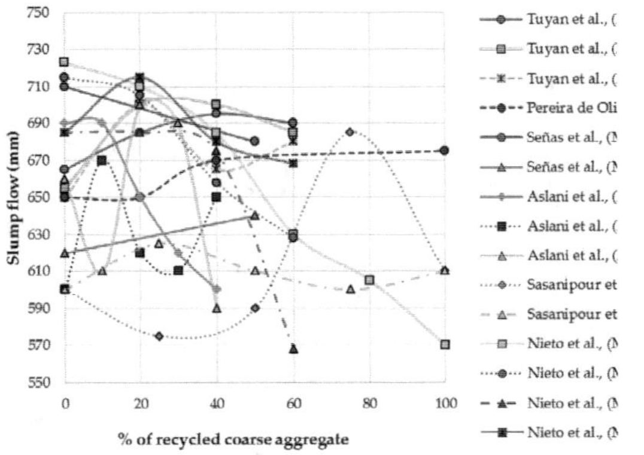

Figure 1. Effect of slump flow on the percentage of recycled coarse aggregate

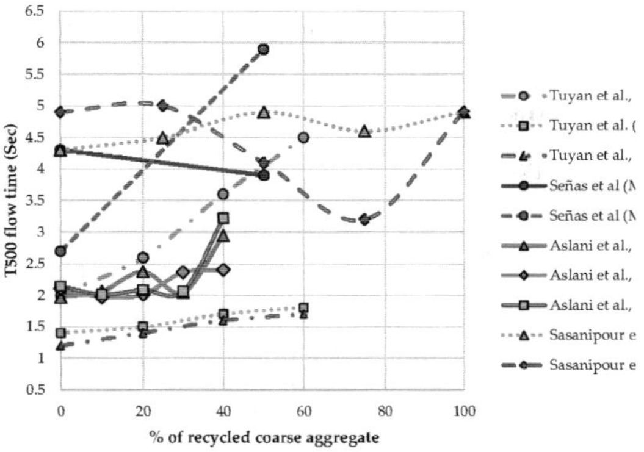

Figure 2. Effect of T500 flow time on the percentage of recycled coarse aggregate.

3.2 Influence of design parameters on Strength

From the available literature, the results are grouped into six families based on the compressive strength grade. Twenty-eight days compressive strength of control or reference SCC mix is used to classify the proportions irrespective of the size of the specimen, shape of the specimen, and various international standards or codes used for testing of the specimen. For example, Family I consist of data based on the reference or control 28 days compressive strength lies in the range of 70 MPa to 80 MPa. In a similar fashion, family II, family III, family IV, family V, and family VI consists of 28 days compressive strength that lies in the range of 60 MPa to 70 MPa, 50 MPa to 60 MPa, 40 MPa to 50 MPa, 30 MPa to 40 MPa, and 20 MPa to 30 MPa, respectively. Corresponding split tensile strength at 28 days is also tabulated irrespective of nature of size, the shape of the specimen, and international standards or codes. Effect of different parameters on the strength is shown in figure 3.

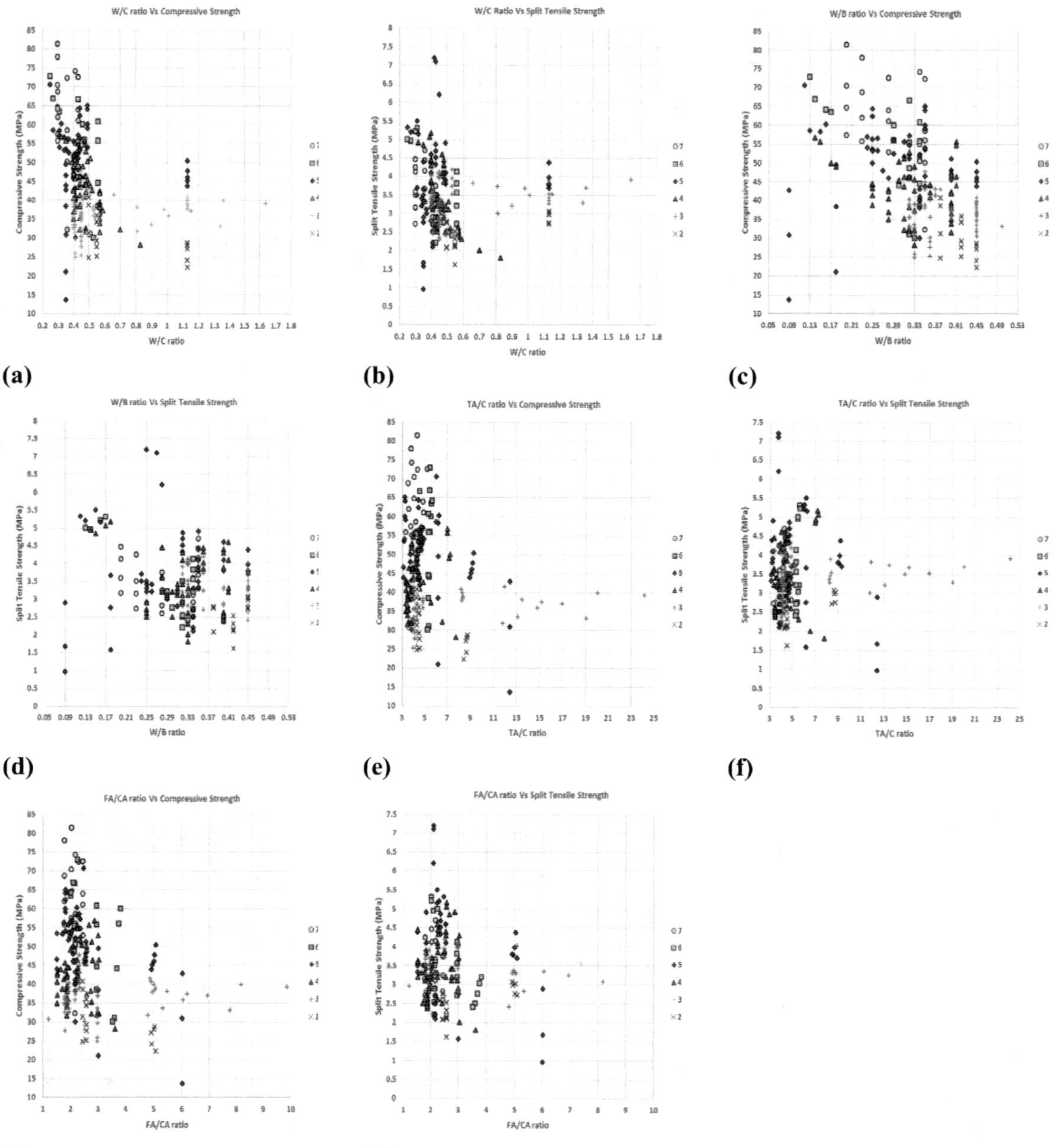

(a) (b) (c) (d) (e) (f) (g) (h)

Figure 3. (a) Effect of W/C ratio on various compressive strengths of SCC concrete with RA according to the literature. (b) Effect of W/C ratio on various split tensile strength of SCC concrete with RA according to the literature. (c) Effect of the W/B ratio on various compressive strengths of SCC concrete with RA according to the literature. (d) Effect of the W/B ratio on various split tensile strength of SCC concrete with RA according to the literature. (e) Effect of TA/C ratio on various compressive strengths of SCC concrete with RA according to the literature. (f) Effect of TA/C ratio on various split tensile strengths of SCC concrete with RA according to the literature. (g) Effect of FA/CA ratio on various compressive strengths of SCC concrete with RA according to the literature. (h) Effect of the FA/CA ratio on various split tensile strengths of SCC concrete with RA according to the literature.

4. Conclusion

An intensive review on the impact of design parameters was conducted in order to make a simplified design procedure for the design mix of SCC made with recycled aggregates. The overall summarized report on the review is stated below:

- Increase in replacement of coarse aggregate by RA results in T500 increases with SF decreases.
- An increase in replacement percentage of natural aggregate by the coarse aggregate result in an increase in flow time.
- Increase in W/C ratio results in a decrease in compressive strength and split tensile strength.
- W/B ratio less than 0.30 produces higher compressive strength and, simultaneously, higher split tensile strength. Further increase in the W/B ratio results in a decrease in compressive strength and split tensile strength.
- The lower the TA/C ratio is, the higher the compressive strength will be.

Detailed investigation on the effect of design parameters on the properties of SCC with recycled aggregate will help researchers and engineers to develop a minimum requirement for a SCC mix with recycled aggregates.

References

[1] Padmini, A. K.; Ramamurthy, K.; Mathews, M. S. Influence of Parent Concrete on the Properties of Recycled Aggregate Concrete. *Constr. Build. Mater.* **2009**, *23* (2), 829–836. https://doi.org/10.1016/j.conbuildmat.2008.03.006.

[2] Singh, A.; Arora, S.; Sharma, V.; Bhardwaj, B. Workability Retention and Strength Development of Self-Compacting Recycled Aggregate Concrete Using Ultrafine Recycled Powders and Silica Fume. *J. Hazardous, Toxic, Radioact. Waste* **2019**, *23* (4), 04019016. https://doi.org/10.1061/(ASCE)HZ.2153-5515.0000456.

[3] Kaza, S.; Yao, L. C.; Bhada-Tata, P.; Van Woerden, F. *What a Waste 2.0: A Global Snapshot of Solid Waste Management to 2050*; Washington, DC: World Bank, 2018. https://doi.org/10.1596/978-1-4648-1329-0.

[4] Suárez-Silgado, S.; Molina, J. D. A.; Mahecha, L.; Calderón, L. Diagnóstico y Propuestas Para La Gestión de Los Residuos de Construcción y Demolición En La Ciudad de Ibagué (Colombia). *Gestión y Ambient.* **2018**, *21* (1), 9–21. https://doi.org/10.15446/ga.v21n1.69637.

[5] Albeiro, C.; Bustos, P.; Guillermo, L.; Pumarejo, F.; Humberto, É.; Cotte, S.; Alexander, H.; Quintana, R. Residuos de Construcción y Demolición (RCD), Una Perspectiva de Aprovechamiento Para La Ciudad de Barranquilla Desde Su Modelo de Gestión. *Ing. y Desarro.* **2017**, *35* (2), 533–555. https://doi.org/10.14482/inde.32.2.5406.

[6] UNEP. *REPORT OF THE UNITED NATIONS ENVIRONMENT PROGRAMME*; New York, 2015.

[7] Robalo, K.; Costa, H.; do Carmo, R.; Júlio, E. Experimental Development of Low Cement Content and Recycled Construction and Demolition Waste Aggregates Concrete. *Constr. Build. Mater.* **2021**, *273*, 121680. https://doi.org/10.1016/j.conbuildmat.2020.121680.

[8] Vaishnavi Devi, S.; Gausikan, R.; Chithambaranathan, S.; Wilfred Jeffrey, J. Utilization of Recycled Aggregate of Construction and Demolition Waste as a Sustainable Material. *Mater. Today Proc.* **2021**. https://doi.org/10.1016/j.matpr.2020.12.013.

[9] Pacheco-Torgal, F. High Tech Startup Creation for Energy Efficient Built Environment. *Renewable and Sustainable Energy Reviews*. Elsevier Ltd 2017, pp 618–629. https://doi.org/10.1016/j.rser.2016.12.088.

[10] García González, J.; Juan Valdés, A.; Morán del Pozo, J. M.; Belie, N. De. Concrete with Ceramic Mixed Construction and Demolition Waste: Optimization of Physical, Mechanical and

Durability Properties by Aggregate Pre-Saturation, Use of Superplasticizers and Microbially Induced Carbonate Precipitation, Universidad de León, 2016.

[11] Mas, B.; Cladera, A.; Olmo, T. del; Pitarch, F. Influence of the Amount of Mixed Recycled Aggregates on the Properties of Concrete for Non-Structural Use. *Constr. Build. Mater.* **2012**, *27* (1), 612–622. https://doi.org/10.1016/j.conbuildmat.2011.06.073.

[12] Kapoor, K.; Singh, S. P.; Singh, B.; Singh, P. Effect of Recycled Aggregates on Fresh and Hardened Properties of Self Compacting Concrete. In *Materials Today: Proceedings*; Elsevier Ltd, 2020; Vol. 32, pp 600–607. https://doi.org/10.1016/j.matpr.2020.02.753.

[13] Fenollera, M.; Míguez, J.; Goicoechea, I.; Lorenzo, J. Experimental Study on Thermal Conductivity of Self-Compacting Concrete with Recycled Aggregate. *Materials (Basel).* **2015**, *8* (7), 4457–4478. https://doi.org/10.3390/ma8074457.

[14] Santos, S. A.; da Silva, P. R.; de Brito, J. Mechanical Performance Evaluation of Self-Compacting Concrete with Fine and Coarse Recycled Aggregates from the Precast Industry. *Materials (Basel).* **2017**, *10* (8). https://doi.org/10.3390/ma10080904.

[15] Uysal, M.; Yilmaz, K. Effect of Mineral Admixtures on Properties of Self-Compacting Concrete. *Cem. Concr. Compos.* **2011**, *33* (7), 771–776. https://doi.org/10.1016/j.cemconcomp.2011.04.005.

[16] Carro-López, D.; González-Fonteboa, B.; De Brito, J.; Martínez-Abella, F.; González-Taboada, I.; Silva, P. Study of the Rheology of Self-Compacting Concrete with Fine Recycled Concrete Aggregates. *Constr. Build. Mater.* **2015**, *96*, 491–501. https://doi.org/10.1016/j.conbuildmat.2015.08.091.

[17] Duan, Z.; Singh, A.; Xiao, J.; Hou, S. Combined Use of Recycled Powder and Recycled Coarse Aggregate Derived from Construction and Demolition Waste in Self-Compacting Concrete. *Constr. Build. Mater.* **2020**, *254*, 119323. https://doi.org/10.1016/j.conbuildmat.2020.119323.

[18] Arafa, M.; Tayeh, B. A.; Alqedra, M.; Shihada, S.; Hanoona, H. Investigating the Effect of Sulfate Attack on Compressive Strength of Recycled Aggregate Concrete. *J. Eng. Res. Technol. JERT* **2017**, *4* (4), 137–143.

[19] Alaejos, P.; De Juan, M. S.; Rueda, J.; Drummond, R.; Valero, I. Quality Assurance of Recycled Aggregates. *RILEM State-of-the-Art Reports* **2013**, *8*, 229–273. https://doi.org/10.1007/978-94-007-4908-5_6.

[20] Panda, K. C.; Bal, P. K. Properties of Self Compacting Concrete Using Recycled Coarse Aggregate. In *Procedia Engineering*; Elsevier Ltd, 2013; Vol. 51, pp 159–164. https://doi.org/10.1016/j.proeng.2013.01.023.

[21] Manzi, S.; Mazzotti, C.; Bignozzi, M. C. Self-Compacting Concrete with Recycled Concrete Aggregate: Study of the Long-Term Properties. *Constr. Build. Mater.* **2017**, *157*, 582–590. https://doi.org/10.1016/j.conbuildmat.2017.09.129.

[22] Mohammed, S. I.; Najim, K. B. Mechanical Strength, Flexural Behavior and Fracture Energy of Recycled Concrete Aggregate Self-Compacting Concrete. *Structures* **2020**, *23*, 34–43. https://doi.org/10.1016/j.istruc.2019.09.010.

[23] Tuyan, M.; Mardani-aghabaglou, A.; Ramyar, K. Freeze – Thaw Resistance , Mechanical and Transport Properties of Self-Consolidating Concrete Incorporating Coarse Recycled Concrete Aggregate. *Mater. Des.* **2014**, *53*, 983–991. https://doi.org/10.1016/j.matdes.2013.07.100.

[24] Pereira-de-Oliveira, L. A.; Nepomuceno, M. C. S.; Castro-Gomes, J. P.; Vila, M. F. C. Permeability Properties of Self-Compacting Concrete with Coarse Recycled Aggregates. *Constr. Build. Mater.* **2014**, *51*, 113–120. https://doi.org/10.1016/J.CONBUILDMAT.2013.10.061.

[25] Señas, L.; Priano, C.; Marfil, S. Influence of Recycled Aggregates on Properties of Self-Consolidating Concretes. *Constr. Build. Mater.* **2016**, *113*, 498–505. https://doi.org/10.1016/J.CONBUILDMAT.2016.03.079.

[26] Tang, W. C.; Ryan, P. C.; Cui, H. Z.; Liao, W. Properties of Self-Compacting Concrete with Recycled Coarse Aggregate. *Adv. Mater. Sci. Eng.* **2016**, *2016*. https://doi.org/10.1155/2016/2761294.

[27] Djelloul, O. K.; Menadi, B.; Wardeh, G.; Kenai, S. Performance of Self-Compacting Concrete Made with Coarse and Fine Recycled Concrete Aggregates and Ground Granulated Blast-Furnace Slag. **2018**, *6* (2), 103–121. https://doi.org/10.12989/acc.2018.6.2.103.
[28] Sasanipour, H.; Aslani, F. Durability Properties Evaluation of Self-Compacting Concrete Prepared with Waste Fine and Coarse Recycled Concrete Aggregates. *Constr. Build. Mater.* **2020**, *236*, 117540. https://doi.org/10.1016/j.conbuildmat.2019.117540.
[29] Aslani, F.; Ma, G.; Yim Wan, D. L.; Muselin, G. Development of High-Performance Self-Compacting Concrete Using Waste Recycled Concrete Aggregates and Rubber Granules. *J. Clean. Prod.* **2018**, *182*, 553–566. https://doi.org/10.1016/j.jclepro.2018.02.074.
[30] Ouldkhaoua, Y.; Benabed, B.; Abousnina, R.; Kadri, E. H.; Khatib, J. Effect of Using Metakaolin as Supplementary Cementitious Material and Recycled CRT Funnel Glass as Fine Aggregate on the Durability of Green Self-Compacting Concrete. *Constr. Build. Mater.* **2020**, *235*, 117802. https://doi.org/10.1016/j.conbuildmat.2019.117802.
[31] Nieto, D.; Dapena, E.; Alaejos, P.; Olmedo, J.; Pérez, D. Properties of Self-Compacting Concrete Prepared with Coarse Recycled Concrete Aggregates and Different Water:Cement Ratios. *J. Mater. Civ. Eng.* **2019**, *31* (2), 04018376. https://doi.org/10.1061/(ASCE)MT.1943-5533.0002566.
[32] Gupta, P. K.; Rajhans, P.; Panda, S. K.; Nayak, S.; Das, S. K. Mix Design Method for Self-Compacting Recycled Aggregate Concrete and Its Microstructural Investigation by Considering Adhered Mortar in Aggregate. *J. Mater. Civ. Eng.* **2020**, *32* (3), 04019371. https://doi.org/10.1061/(ASCE)MT.1943-5533.0003014.
[33] Abed, M.; Nemes, R.; Lublóy, É. Performance of Self-Compacting High-Performance Concrete Produced with Waste Materials after Exposure to Elevated Temperature. *J. Mater. Civ. Eng.* **2020**, *32* (1), 05019004. https://doi.org/10.1061/(ASCE)MT.1943-5533.0002989.
[34] Xavier, B. C.; Verzegnassi, E.; Bortolozo, A. D.; Alves, S. M.; Cecche Lintz, R. C.; Andreia Gachet, L.; Osório, W. R. Fresh and Hardened States of Distinctive Self-Compacting Concrete with Marble- and Phyllite-Powder Aggregate Contents. *J. Mater. Civ. Eng.* **2020**, *32* (5), 04020065. https://doi.org/10.1061/(ASCE)MT.1943-5533.0003103.
[35] Dapena, E.; Alaejos, P.; Lobet, A.; Pérez, D. Effect of Recycled Sand Content on Characteristics of Mortars and Concretes. *J. Mater. Civ. Eng.* **2011**, *23* (4), 414–422. https://doi.org/10.1061/(ASCE)MT.1943-5533.0000183.
[36] Rajhans, P.; Gupta, P. K.; Ranjan, R. K.; Panda, S. K.; Nayak, S. EMV Mix Design Method for Preparing Sustainable Self Compacting Recycled Aggregate Concrete Subjected to Chloride Environment. *Constr. Build. Mater.* **2019**, *199*, 705–716. https://doi.org/10.1016/j.conbuildmat.2018.12.079.
[37] Safiuddin, M.; Salam, M. A.; Jumaat, M. Z. Effects of Recycled Concrete Aggregate on the Fresh Properties of Self-Consolidating Concrete. *Arch. Civ. Mech. Eng.* **2011**, *11* (4), 1023–1041. https://doi.org/10.1016/S1644-9665(12)60093-4.
[38] Grdic, Z. J.; Toplicic-Curcic, G. A.; Despotovic, I. M.; Ristic, N. S. Properties of Self-Compacting Concrete Prepared with Coarse Recycled Concrete Aggregate. *Constr. Build. Mater.* **2010**, *24* (7), 1129-1133[1] Grdic, Z. J.; Toplicic-Curcic, G. A.; https://doi.org/10.1016/j.conbuildmat.2009.12.029.

IOP Conf. Series: Materials Science and Engineering 1222 (2022) 012013 doi:10.1088/1757-899X/1222/1/012013

Performance Comparison of Variants Based on Swarm Intelligence Algorithm of Mathematical and Structural Optimization

Phuc Tran Van[1,a], Thu Huynh Van[2,b], Sawekchai Tangaramvong[2,c]

[1] University of Architecture Ho Chi Minh City, Pasteur str., 196, Ho Chi Minh, 700000, Vietnam
[2]Applied Mechanics and Structures Research Unit, Department of Civil Engineering, Chulalongkorn University, 10330 Bangkok, Thailand

[a]phuc.tranvan@uah.edu.vn, [b]6278306421@student.chula.ac.th, [c]sawekchai.t@chula.ac.th

Abstract. In this paper, three different variants of the canonical particle swarm optimization (PSO) algorithm are compared in terms of their optimal solutions to engineering optimization problems. The PSO often face to premature convergence due to its weak capability in local exploitation, which leads to a low optimization precision or even failure. Moreover, the computational burden is also one of the obstacles for the PSO when solving complex problems. To overcome these drawbacks, the original PSO algorithm is integrated with the knowledge sharing and auxiliary mechanisms of comprehensive learning (CL) and Gaussian local search (GLS) strategies to form the so-called GLS-PSO, CLPSO, and GLS-CLPSO algorithms. To demonstrate the accuracy and robustness of the three variants, their performance is tested on a benchmark spatial truss structure with continuous variables. The obtained optimization results are then compared with those by other PSO variants in the literature.

1. Introduction

Optimization problems are commonly found in science and engineering applications and they are becoming more and more complicated. Some of well-known traditional approaches to solving these problems have efficiently provided theoretical concepts for the ground understanding but also solution techniques. However, some drawbacks of these techniques still exist, such as their gradient-based formulations as well as complexity in calculation of the derivatives and sensitivity of the objective and constraint functions at the starting point. Therefore, most of the traditional techniques are unfit for solving complex optimization problems. Recently, swarm intelligence (SI) algorithms belong to the stochastic search techniques which have gained popularity as an alternative solution for the traditional techniques. One of the famous SI algorithms is the particle swarm optimization (PSO) that was firstly developed by Kennedy and Eberhart [1]. The algorithm is constructed based on the intelligent behavior of animals (e.g., the colony of fish and birds) in finding food sources and avoiding enemies in nature. Due to the simplicity of implementation, and requires only a few parameters, the PSO has become the most attractive algorithm and successfully applied for many real applications [2]. However, like other stochastic search techniques, the crude PSO often suffers from trapping into local optima, with premature convergence but less accuracy. This is likely faced when solving complex problems and has attracted the attention of many researchers to enhance the performance of the crude PSO. Many variant and hybrid versions have proposed either modifying the tuning parameters, solution update formula, and auxiliary mechanisms or combining with other existing algorithms [3].

In this study, three variants of the crude PSO algorithm are proposed by integrating with comprehensive learning (CL) and Gaussian local search (GLS) strategies as respectively termed GLS-PSO, CLPSO, and GLS-CLPSO algorithms. In particular, the CL strategy is used to improve the *exploration capability* of the PSO, here exemplars are constructed by utilizing personal best experiences of the entire population during the process of updated positions. Meanwhile, the GLS strategy performs a neighborhood search that uses the Gaussian distribution to generate samples through the information sharing of personal best and global best positions from the PSO. This strategy helps the PSO to strongly improve its *exploitation capability*. The accuracy and effectiveness of the proposed algorithm are illustrated through a spatial truss design benchmark. The obtained optimization results were then compared with those reported in the literature.

2. Formulation of the optimization problem

Generally, a design optimization problem is formulated for minimizing (or maximizing) an objective function $f(\mathbf{A})$, while satisfying certain design constraints, such that

$$\text{Minimize } f(\mathbf{A}), \quad \mathbf{A} = [x_1, x_2, \ldots, x_{nd}]. \tag{1}$$

$$\text{Subjected to } g_j(\mathbf{A}) \leq 0, \quad j = 1, 2, \ldots, nc. \tag{2}$$

where $\mathbf{A} \in \Re^{nd}$ denotes the vector of nd-dimensional design variables, each element of $\mathbf{A}$ is defined by a lower bound $x_{g,\min}$ and an upper bound $x_{g,\max}$ (g = 1,.., nd), and $g_j(\mathbf{A})$ represents the j-th constraint among a total of nc constraints.

3. Comprehensive learning PSO

Consider a swarm of a finite number of particles. Let *pbest* and *gbest* denote the best position each particle has seen and the best position found by any particle, respectively. In the crude PSO, each particle mimics the moving trajectory of *pbest* and *gbest* simultaneously. The *gbest* is a piece of important information to navigate the direction of the whole swarm. However, when particles in the swarm learn from *gbest* even if the current *gbest* is far from the global solution. This results in easily attracting to the region around *gbest* and trapping into a local minimum. In order to improve the sharing of information, Liang and Huang [4] developed the CLPSO algorithm using the CL strategy. The CL uses *pbest* from all particles to maintain the population diversity as well as enhancing the exploration capability. Each particle learns from different exemplars on each dimensions. The next velocity and position of CLPSO are updated according to Eqs. (3) and (4).

$$V_i^{next} = w \times V_i + c_1 \times r_1 \times \left(pbest_{fi(D)} - X_i\right) + c_2 \times r_2 \times \left(gbest - X_i\right). \tag{3}$$

$$X_i^{next} = X_i + V_i^{next}. \tag{4}$$

where fi = [$fi(1)$, $fi(2)$,..., $fi(nd)$] denotes if the i-th particle follows its own or other's *pbest* for each dimension nd, X_i the position of particle i, V_i the velocity of particle i, w an inertial weight scaling factor, c_1 and c_2 two acceleration coefficients scaling respectively the influences of the *pbest* and *gbest*, and r_1 and r_2 independent random numbers within the interval of [0,1], $pbest_{fi(nd)}$ denotes the exemplar of the i-th particle on the nd-dimension, *gbest* the best position found by the entire swarm as follows in two Eqs. (5) and (6), respectively.

$$pbest_{fi(nd)}^{next} = \begin{cases} pbest_{fi(nd)} & if \;\; f\left(pbest_{fi(nd)}\right) \leq f\left(X_i\right) \\ X_i & if \;\; f\left(pbest_{fi(nd)}\right) > f\left(X_i\right) \end{cases} \quad for \; \forall i \in \{1,? \;\; ps\} \tag{5}$$

$$gbest = \min\left\{f\left(pbest_{fi(nd)}\right), f\left(gbest\right)\right\} \quad for\ \forall i \in \{1,\ldots,ps\} \tag{6}$$

The CL strategy is illustrated as follows [5]:

Step 1: If the *i*-th particle's historical *pbest* does not improve consecutively and over a refreshing gap *m* (set *m* = 7), then two particles are randomly choosed from the population (population size *ps*).

Step 2: The objective function values of these two particles are compared together, and the better one is selected as the exemplar based on the learning probability Pc_i, that is determined by Eq. (7). The process is repeated for all dimensions to yield the new direction $pbest_{fi(nd)}$ of the *i*-th particle.

Step 3: If the exemplar located the same position with *pbest*, then a dimension is randomly selected from another particle's *pbest* in the corresponding dimension.

$$Pc_i = 0.05 + 0.45\left(\exp\left(\frac{10(i\text{-}1)}{ps\text{-}1}\right) - 1\right) / \left(\exp(10) - 1\right). \tag{7}$$

4. Gaussian local search

Unlike other distributions, the Gaussian distribution describes a bell-curved shape. It exploits over a central filed, and improves the optimization accuracy. The distribution typically denoted by $\mathcal{N}(\mu, \sigma^2)$ is characterized a mean value μ and variance σ^2. The Gaussian probability density function is defined by:

$$f(x) = \frac{1}{\sqrt{2\pi\sigma^2}}\exp\left(-\frac{(x-\mu)^2}{2\sigma^2}\right), \quad \sigma > 0. \tag{8}$$

The GLS strategy improves the exploitation ability using the particle *gbest* and *pbest* from the PSO procedure. The random numbers around *gbest* are generated by Gaussian distribution [6]. To characterize Gaussian distribution and obtain the random values, the mean and standard deviation are determined by Eq. (9):

$$x_i \cong \mathcal{N}\left(\mu, \sigma^2\right) = \mathcal{N}\left(gbest, \left|gbest - pbest_i\right|\right). \tag{9}$$

The number of GLS iterations is reset such that the *gbest* exploits sufficiently the current region and jumps out of the local optima.

5. Illustrative Example

Consider a space 120-bar dome truss as shown in Fig 1 [7, 8]. Due to symmetry of the dome structure geometry, all 120 members are divided into 7 different groups (*nd* = 7). The members of each group have the same length and cross-sectional area. In addition, the allowable tensile and compressive stress are defined complying with the AISC-ASD (1989) [9] specification, as follows:

$$\begin{cases} \sigma_m^+ = 0.6F_y & for\ \sigma_m \geq 0 \\ \sigma_m^- & for\ \sigma_m < 0 \end{cases} \tag{10}$$

Where σ_m^- is calculated based on the slenderness ratio

$$\sigma_m^- = \begin{cases} \left[\left(1 - \frac{\lambda_m^2}{2C_c^2}\right)Fy\right] / \left(\frac{5}{3} + \frac{3\lambda_m}{8C_c} - \frac{\lambda_m^3}{8C_c^3}\right) & if\ \lambda_m < C_c \\ \frac{12\pi^2 E}{23\lambda_m^2} \quad if\ \lambda_m \geq C_c \end{cases} \tag{11}$$

where yield stress of F_y = 58.0 ksi (400 Mpa), C_c is the member slenderness limit ($C_c = \sqrt{2\pi^2 E / F_y}$), λ_m is the slenderness ratio of member m ($\lambda_m = k_m l_m / r_m$), k_m is the effective length factor (k_m = 1 for all members), r_m is the radius of gyration that can be expressed based on the cross-sectional areas as $r_m = kA_m^p$, in which k and p are constants related to the types of cross sections for truss members (e.g., pipes, angles, and tees). In this example, pipe sections (k = 0.4993 and p = 0.6777) are considered for bars [7].

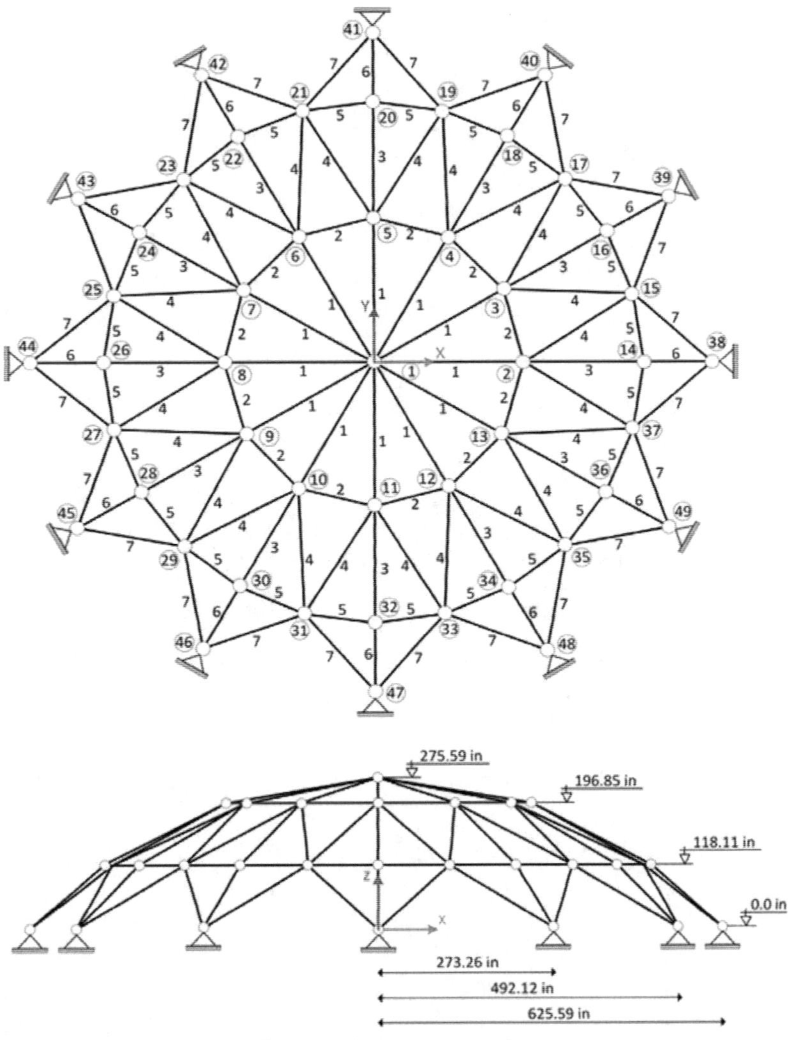

Figure 1. Space 120-bar dome truss structure

Young's modulus and density of material are 30,450 ksi (210,000 Mpa) and 0.288 lb/in^3 (7971.81 kg/m^3), respectively. The displacement limitations of ± 0.1969 (5 mm) are imposed at all nodes in x, y, and z directions. The dome truss is subjected to vertical forces at all the unsupported nodes. These forces involve -13.49 kips (-60 kN) at node 1, -6.744 kips (-30 kN) at nodes 2 through 14, and -2.248 kips (-10 kN) at the rest of the nodes. The minimum and maximum available areas of all truss members are specified as 0.775 in^2 (5 cm^2) and 20.0 in^2 (129.032 cm^2), respectively. Three proposed PSO variants are encoded in Python, and run by Intel Core i5-9400 CPU @ 2.9 GHz.

For the three proposed PSO variants, a population of 30 particles is used, c_1 and c_2 are set equal to 2.0, and the inertia weight w is calculated by linearly decreasing from 0.9 to 0.2. The stopping criterion is assigned as the maximum number of iterations, set equal to 1500. The number of GLS iterations is 1000.

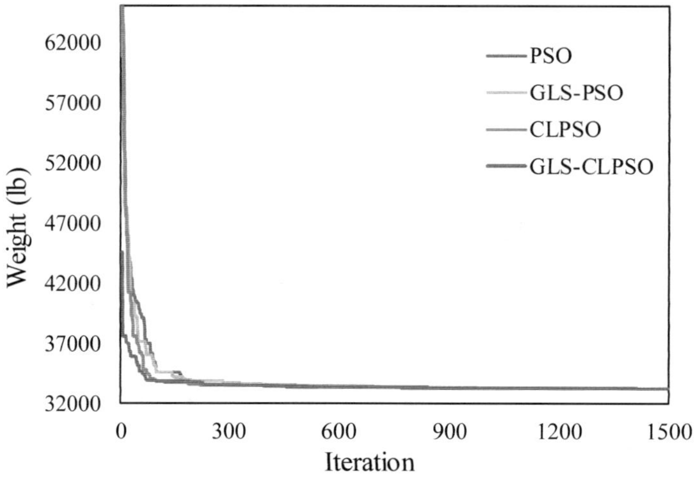

Figure 2. Solution convergences

As a result, Fig. 2 reports the weight convergence of the truss structure by the PSO and the three proposed variants. Table 1 provides the performance comparison of three proposed PSO variants with the original PSO and other PSO variants in the literature. Noticeably, the GLS-CLPSO provides outstanding results without violating the limitations of the maximum displacement and stresses. It is also seen that the minimum weight of 120-bar dome by the GLS-CLPSO is 33,250.66 lb, which is lighter than most designs by CLPSO, GLS-PSO, PSO, HPSSO, and MSPSO.

Table 1. Performance comparison of optimal solutions of 120-bar truss design.

Group	Literature		Present study			
	MSPSO [10]	HPSSO [7]	PSO	GLS-PSO	CLPSO	GLS-CLPSO
	x_i (in^2)	x_i (in^2)	x_i (in^2)	x_i (in^2)	x_i (in^2)	x_i (in^2)
x_1	3.0244	3.0241	3.0277	3.0287	3.0309	3.0274
x_2	14.7804	14.7808	14.0977	14.4397	14.6184	15.2985
x_3	5.0567	5.0522	5.1552	5.2001	5.3269	5.0303
x_4	3.1359	3.1369	3.1066	3.0919	3.0880	3.0733
x_5	8.4830	8.5004	8.7257	8.5107	8.3481	8.2674
x_6	3.3104	3.2888	3.4711	3.4527	3.4091	3.4077
x_7	2.4977	2.4969	2.4963	2.5057	2.4969	2.5010
Best weight (lb)	33,251.22	33,250.05	**33,260.82**	**33,252.31**	**33,251.78**	**33,250.66**

In summary, the CLPSO is apparently good at preserving the particles' diversity and is thus excellent in exploration. However, it is weak in the exploitation ability. Meanwhile, the GPSO is efficient in performing a local search, but weak in the exploration ability. The GLS-CLPSO may be a remedy that significantly improves the exploitation ability of the CLPSO and the exploration ability of the GPSO.

6. Conclusions

This paper proposed three variants of the PSO algorithm by integrating it with CL and GLS strategies, namely, GLS-PSO, CLPSO, and GLS-CLPSO. To demonstrate the effectiveness of these variants, their performance was tested on a space truss structure. The optimal results showed that the GLS-CLPSO algorithm outperforms the PSO, CLPSO, GLS-CLPSO, and other PSO variants in the literature. Finally, it can be concluded that the PSO variants provide suitable ways to achieve not only a good balance between exploration and exploitation but also a competitive solution to solve complicated optimization problems in reality.

References

[1] Eberhart R, Kennedy J. (1995) Particle swarm optimization, *proceeding of IEEE International Conference on Neural Network*. Perth, Australia. pp. 1942-8.
[2] Mavrovouniotis M, Li C, Yang S. (2017) A survey of swarm intelligence for dynamic optimization: Algorithms and applications. *Swarm and Evolutionary Computation*. **33** pp. 1-17.
[3] Wang D, Tan D, Liu L. (2018) Particle swarm optimization algorithm: an overview. *Soft Computing*. **22** pp. 387-408.
[4] Liang JJ, Qin AK, Suganthan PN, Baskar S. 2006 Comprehensive learning particle swarm optimizer for global optimization of multimodal functions. *IEEE Transactions on Evolutionary Computation*. **10** pp. 281-95.
[5] Zhang K, Huang Q, Zhang Y. (2019) Enhancing comprehensive learning particle swarm optimization with local optima topology. *Information Sciences*. **471** pp. 1-18.
[6] Jia D, Zheng G, Qu B, Khan MK. (2011) A hybrid particle swarm optimization algorithm for high-dimensional problems. *Computers & Industrial Engineering*. **61** pp. 1117-22.
[7] Kaveh A, Bakhshpoori T, Afshari E. (2014) An efficient hybrid particle swarm and swallow swarm optimization algorithm. *Computers and Structures*. **143** pp. 40-59.
[8] Jafari M, Salajegheh E, Salajegheh J. (2019) An efficient hybrid of elephant herding optimization and cultural algorithm for optimal design of trusses. *Engineering with Computers*. **35** pp. 781-801.
[9] AISC-Manual of steel construction–allowable stress design. American Institute of Steel Construction, Chicago, 1989.
[10] Talatahari S, Kheirollahi M, Farahmandpour C, Gandomi AH. (2013) A multi-stage particle swarm for optimum design of truss structures. *Neural Computing and Applications*. **23** pp. 1297-309.

Modified Scheil equation for the Cu-Sn system and comparison of hardness and electrical conductivity of Cu-8.5wt% Sn alloy directionally solidified

R A Cruz[1, 2], G A Santos[1], M S Nascimento[1], R Teram[1, 2], R de Luca[1,4], V T dos Santos[2,4], M R da Silva[4] and A A Couto[2, 3]

[1] Federal Institute of Education, Science and Technology of São Paulo, São Paulo, SP, Brazil
[2] Mackenzie Presbyterian University, São Paulo, SP, Brazil
[3] Institute of Nuclear and Energy Research, IPEN, São Paulo, SP, Brazil
[4]Termomecanica São Paulo S.A., São Bernardo do Campo, SP, Brazil

ricardoapcruz@ifsp.edu.br

Abstract. The Scheil equation, which estimates the concentration of solute in the solid at the solid/liquid interface, does not account for the curvatures of *liquidus* and *solidus* lines. We modified the equation to account for these curvatures and compared some microsegregation indexes obtained by the original and the new equations with data from the literature about the Cu-Sn system. The new one can furnish a better approximation of the range of concentration of solute, but a poorer estimate of the maximum volume fraction of the eutectoid mixture. At last, we compared the hardness and electrical conductivity of an upward solidified alloy of this system with its equivalent commercial one. The as-cast alloy has a lower hardness and a higher conductivity due to its columnar structure.

1. Introduction

Microsegregation is a physical-chemical phenomenon. It occurs at the microstructural level during the solidification of an alloy. The difference in solubility for solute elements in the solid phase and the liquid phase drives that phenomenon. Under most solidification conditions, equilibrium can exist at the solid/liquid interface, and the concentrations of solid, C_S, and liquid, C_L, for any given element at the interface, can be used to define a partition ratio, k:

$$k = \frac{C_L}{C_{Si}} \tag{1}$$

C is concentration; subscripts S and L indicate solid and liquid, respectively; and subscript i indicates a quantity at the solid/liquid interface.

The Scheil equation estimates the concentration of solute in the solid at the solid/liquid interface. It assumes complete diffusion of solute in the liquid and zero diffusion in the solid, and can be written as follows [1].

$$C_{Si} = kC_0(1 - f_S)^{k-1} \tag{2}$$

IOP Conf. Series: Materials Science and Engineering 1222 (2022) 012014 doi:10.1088/1757-899X/1222/1/012014

C_0 is overall alloy concentration, and f_S is volume fraction of solid.

The partition ratio k seems to be not constant since the *liquidus* and *solidus* lines of the Cu-Sn phase diagram are not straight lines. It is necessary to correct this distortion, and we propose a new model to modify Scheil equations appropriately.

We will compare the original Scheil equations with the new ones by estimating some microsegregation indexes, such as the maximum volume fraction of the eutectoid mixture for three different alloys of the Cu-Sn system. We will also compare the hardness and electrical conductivity of an alloy of this system (Cu-8.5wt%Sn) subjected to upward solidification with the equivalent commercial one.

2. Proposed Model

Exists a quadratic function that can suitably relate C_{Si} to C_L and the curvature of the *solidus* line to the *liquidus* line. Using three pairs of points at different temperatures, and solving a system of three equations, we get [2]:

$$C_{Si} = 1.5986 C_L^{\,2} + 0.1218 C_L \tag{3}$$

Equating the amount of solute that passed from the solid into the liquid, we get:

$$(C_L - C_{Si})\mathrm{d}f_S = (1 - f_S)\mathrm{d}C_L \tag{4}$$

Supposing that an increase in the volume of solid implies an equal decrease in liquid ($\mathrm{d}f_S = -\mathrm{d}f_L$), the solving of equations (3) and (4) for the concentration of solute in the liquid phase will give:

$$C_L = \frac{-0.8782 C_0}{1.5986 C_0((1 - f_S)^{0.8782} - 1) - 0.8782(1 - f_S)^{0.8782}} \tag{5}$$

If we substitute (5) in (3), we will get the concentration of solute in the solid interface S/L. See in figure 1 the representation of Scheil equations and the new model equations for concentration of solute (Sn) in liquid and in solid interface as functions of the solid volume fraction (C_0 = 8wt% Sn).

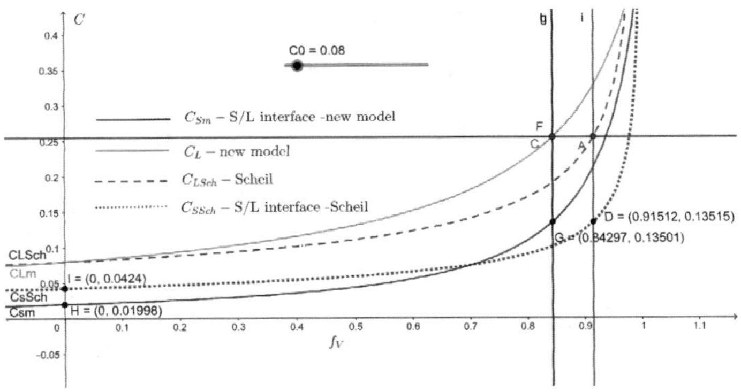

Figure 1. Representation of Scheil equations, (1) and (3), and of the new model equations, (3) and (5)

The area over the Cs function in the interval $0 \leq f_S \leq 1$ is equal to C_0 as to Scheil as to the new model, what indicates that there is mass conservation in both models.

3. Materials and Methods

We made the Cu-8.5wt% Sn alloy with commercially pure tin and UNS C50700 bronze. The melting of the material occurred in an electric furnace under an argon gas atmosphere. The solidification of the alloy occurred inside an upwardly unidirectional solidifying device. Thermocouples, a data acquisition system, and a computer allowed monitoring the temperature in eight sites of the cylindrical casting. More details of this technique is found in the literature [2-4].

We have cut the as-cast cylinder into samples. The sample destined for macrograph (half of the cylindrical casting from a longitudinal cut) was ground, polished, and etched with an acidic ferric chloride solution. Later, we measured electrical conductivity at five different distances from the base of that sample. The central part of the other half of the cylindrical casting gave rise to twelve transverse and twelve longitudinal samples taken from different distances from the base. They were embedded in pairs resulting in six bakelite disks for micrograph and six other disks for chemical analysis and hardness testing.

An inverted platinum microscope was used in conjunction with its software for digital image processing and analysis to make the necessary micrographs. We measured the primary and secondary dendrite arm spacing on the micrographs and proceeded to the statistical analysis of the distribution. The Manual Point Count Method allowed determining the volume fraction of the eutectoid mixture in each transverse sample [2,5,6].

We performed the Vickers hardness testing (HV10) on the six transverse and six longitudinal samples. Optical Emission Spectroscopy enabled the chemical analysis of the transverse samples.

We used a DC10 electrical conductivity meter in a Wheatstone bridge circuit to measure the electrical conductivity in % IACS (International Annealed Copper Standard), according to the ASTM B193 standard.

4. Results and Discussion

The solidification produced a cylindrical casting about 115 mm high after deburring and cleaning. The macrostructure of this casting presented a columnar zone of about 80 mm in length. The analysis of the chemical composition of the transverse samples showed that there was no significant macrosegregation of tin along the height of the casting. The average mass fraction of tin was (8.56 ± 0.05)wt% Sn.

4.1. Hardness HV10, dendrite arm spacing and cooling rate

Figure 2 shows the data of hardness in transversal and longitudinal samples plotted against the secondary dendrite arm spacing. Figure 3 displays the dendrite arm spacing plotted against the cooling rate. We note that the hardness is greater at sites with narrower secondary dendrite arm spacing.

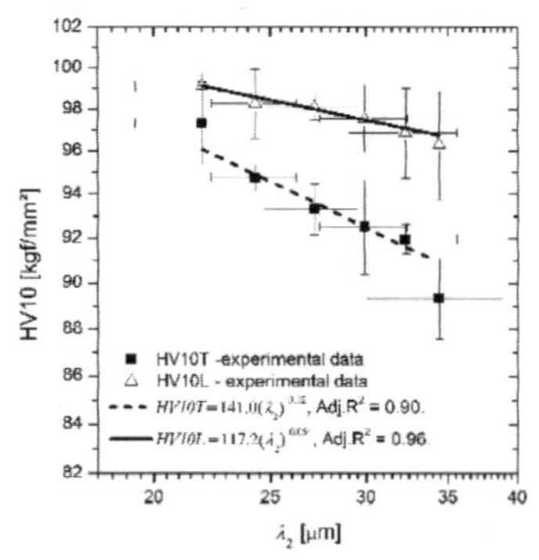

Figure 2. Vickers Hardness against secondary dendrite spacing

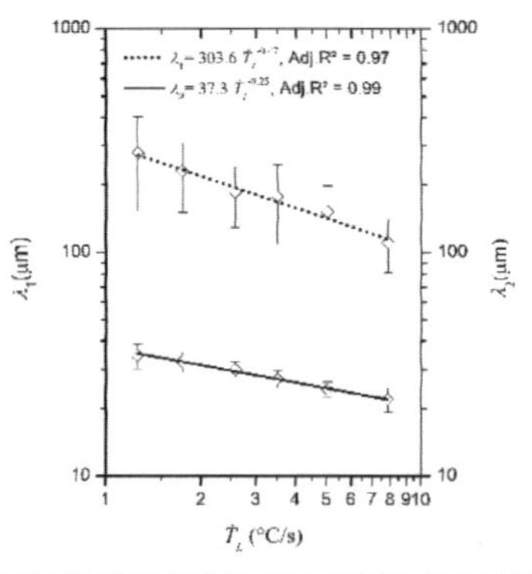

Figure 3. Primary and secondary dendrite arm spacing against cooling rate

A commercial alloy with 8.5%wt Sn (4.9%at Sn) would have a hardness of 128 HV for a grain size of 35 μm (OS035). See figure 4 to check the hardness of commercial alloys as a function of the atomic percentage of tin. The columnar morphology of the alloy obtained in the experiment produced a softer material in comparison with the commercial alloys, the maximum hardness value (98 HV10) is smaller than 128 HV (commercial).

4.2. Electrical conductivity

The electrical conductivity of the material after the solidification process is the same in any site. After five readings, we got the mean value: 14.4±0.3 % IACS.

A commercial alloy with 8.5%wt Sn (4.9%at Sn) would have a resistivity of 139 nΩm. See figure 5 to estimate the resistivity of commercial alloys. Therefore the corresponding conductivity of the commercial alloy is 12.4 % IACS. The columnar morphology of the alloy obtained in the experiment is responsible for a material with better conductivity (an increase of 16%) in comparison with the commercial alloys.

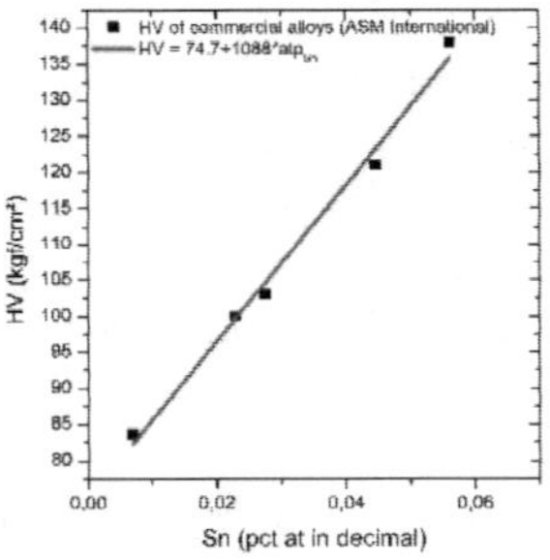

Figure 4. Estimated Vickers Hardness against %at of Sn for commercial alloys - OS035 - data from ASM International [7].

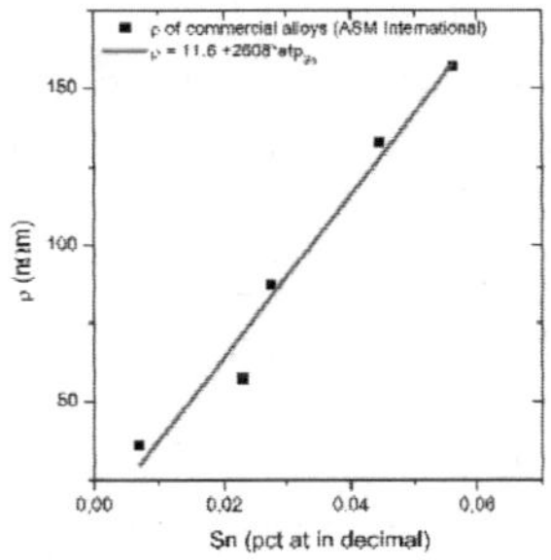

Figure 5. Electrical resistivity against %at of Sn for commercial alloys data from ASM International [7].

4.3. Volume fraction of the eutectoid mixture

The volume fraction of the eutectoid mixture varies from 0.9% to 2.6%. The mean value is 1.5%, and the standard deviation is 0.6%.

We found in the literature that the columnar morphology is associated with low values of volume fraction of eutectoid mixture in comparison to equiaxed morphology [6]. The theoretical maximum value for Vv in Cu-8.5%wtSn alloy is 9.7% (Scheil), and 17% (New model). Both values are by far superior in comparison with the maximum 2.6% measured. One can obtain the theoretical maximum value in this way: During the solidification of an alloy of the Cu-Sn system, the liquid becomes richer in Sn. When the concentration of solute in the liquid reaches 0.255 (25.5%) at the peritectic point, the reminiscent liquid will form a eutectoid mixture. The C_L equation can furnish us with the value of f_S when $C_L = 0.255$. If we subtract this value from 1, we will get the theoretical maximum value of the eutectoid mixture.

Table 5 displays some microsegregation data from the literature in comparison with the predictions of Scheil equation and the ones of the new model (NM) equation. The C_S range can be found substituting $f_S = 0$ and f_S of the peritectic point in C_{Si} equations, (2), (3) and (5).

Table 1. Comparison of some experimental data with the predictions of Scheil equations and of the new model (NM) equations.

	C_0 (%wt)	Cs Exp. Range (%wt)	Cs Scheil Range (%wt)	Cs NM Range (%wt)	Vv exp. Range (%)	Scheil Vv (%)	NM Vv (%)	Observed Morphology
Ref. [6]	8	1.7-14.7	4.2-13.5	2-13.5	0.21-10.8	9.5	15.7	Col. [a], eq. [b]
Ref. [8]	10	2-16	5.3-13.5	2.8-13.5	6-10.9	13.6	21.3	Eq.
Ref. [9]	20	6.5-13	10.6-13.5	8.8-13.5	20 (max)	59.7	62.4	Col.

[a] Columnar
[b] Equiaxed

We can observe that NM equations give a better estimate of the Cs range, while Scheil equations provide a better estimate of the maximum volume fraction of the eutectoid mixture.

5. Conclusions

We can draw the following conclusions:

(1) The modified Scheil equations provide a better estimate of the range of the mass fraction of solute in the solid and a poorer estimate of the maximum volume fraction of the eutectoid mixture.
(2) The upward solidified alloy, with columnar structure, presented better conductivity (an increase of 16%) and lower hardness when compared to the equivalent commercial alloy.
(3) The upward solidified alloy, with a columnar structure, presented a low value of eutectoid mixture in comparison with the estimates of the original Scheil equations and the new ones.

Acknowledgments
The authors are grateful for the support provided by CAPES, Salvador Arena Foundation Educational Center, Termomecanica São Paulo S.A., Mackenzie Presbyterian University, Nuclear and Energy Research Institute, and Federal Institute of São Paulo (IFSP) for the research activities.

References

[1] Smith R. Microsegregation Measurement: Methods and Applications. Metall Mater Trans B Process Metall Mater Process Sci [Internet]. 2018;49(6):3258–79. Available at: https://doi.org/10.1007/s11663-018-1395-4

[2] Cruz RA, Santos GA, Nascimento MS, Frajuca C, Nakamoto FY, da Silva MR, et al. Microstructural characterization and mathematical modeling for determination of volume fraction of eutectoid mixture of the cu-8.5wt% sn alloy obtained by unidirectional upward solidification. Mater Sci Forum. 2020;1012 MSF:302–7. Available at: https://doi.org/10.4028/www.scientific.net/MSF.1012.302

[3] Nascimento M S, dos Santos G A, Teram R, dos Santos V T, da Silva M R, Couto A A. Effects of thermal variables of solidification on the microstructure, hardness, and microhardness of Cu-Al-Ni-Fe alloys. Materials (Basel). 2019;12(8). Available at: https://doi.org/10.3390/ma12081267

[4] Nascimento MS, Tadeu ATR, Frajuca C, Yastami FY, dos Santos GA, Couto AA. An Experimental Study of the Solidification Thermal Parameters Influence upon Microstructure and Mechanical Properties of Al-Si-Cu Alloys. Materials Research 2018;21(5). Available at: https://doi.org/10.1590/1980-5373-MR-2017-0864

[5] Kumoto E A, Alhadeff R O, Martorano M A. Microsegregation and dendrite arm coarsening in tin bronze. Mater Sci Technol. 2002;18(9):1001–6.

[6] Martorano M de A, Capocchi J D T. Effects of Processing Variables on the Microsegregation of Directionally Cast Samples. Mettallurgical Mater Trans A. 31A:3137–48.

[7] ASM International. Metals HandBook VOL 2, Properties and Selection: Nonferrous Alloys and Special-Purpose Materials. 1990.

[8] Kumoto EA, Alhadeff R O, Martorano MA. Microsegregation and dendrite arm coarsening in tin bronze. Mater Sci Technol. 2002;18:1001–6.

[9] Baptista LADS, Paradela KG, Ferreira IL, Garcia A, Ferreira AF. Experimental study of the evolution of tertiary dendritic arms and microsegregation in directionally solidified Al-Si-Cu alloys castings. J Mater Res Technol [Internet]. 2019;8(1):1515–21. Available at: https://doi.org/10.1016/j.jmrt.2018.05.021

AUTHOR INDEX

Institute of Physics
Dirac House, Temple Back
Bristol BS1 6BE UK

ISSN: 1757-8981
ISBN 978-1-7138-4680-2

9 781713 846802

CURRAN ASSOCIATES INC.
proceedings.com